FIXING CLIMATE

FIXING CLIMATE

What Past Climate Changes Reveal About the Current Threat—and How to Counter It

WALLACE S. BROECKER and ROBERT KUNZIG

HILL and WANG
A division of Farrar, Straus and Giroux
New York

Hill and Wang
A division of Farrar, Straus and Giroux
18 West 18th Street, New York 10011

Distributed in Canada by Douglas & McIntyre Ltd.
Printed in the United States of America
First edition, 2008

Library of Congress Cataloging-in-Publication Data
Broecker, Wallace S., 1931–
Fixing climate : what past climate changes reveal about the current threat—and how to counter it / Wallace S. Broecker and Robert Kunzig. — 1st ed.
p. cm.
Includes bibliographical references and index.
ISBN-13: 978-0-8090-4501-3 (hardcover : alk. paper)
ISBN-10: 0-8090-4501-X (hardcover : alk. paper)
1. Climatic changes—History. 2. Global warming. I. Kunzig, Robert. II. Title.

QC981.8.C5B738 2008
551.6—dc22

2008004445

Designed by Jonathan D. Lippincott

www.fsgbooks.com

10 9 8 7 6 5 4 3 2 1

For Gary Comer

Friend, philanthropist, and champion of climate research

Contents

Preface: Taming the Beast

One morning in the spring of 2002, a letter came in with the pile of e-mail that Wally Broecker's assistant prints out for him each day. It was a real paper letter, delivered by the U.S. Postal Service, from someone who, like Broecker, was old enough to consider letters a normal means of communication. The writer's name was Gary Comer, and as Broecker soon learned, he was the founder of Lands' End, the mail-order clothing company. He was also, it seemed, a passionate yachtsman, particularly in Arctic waters, which is how he came to be writing to a climate scientist.

The previous summer, Comer explained, he had been cruising off Greenland on his 151-foot motor yacht, *Turmoil.* Almost on a lark, he and his shipmates had decided to see if they could navigate the Northwest Passage, along the northern coast of Canada all the way to Alaska. European explorers, Comer knew, had spent four centuries trying to find that seaway through the Arctic ice, before Roald Amundsen finally succeeded in 1906. The most notorious of many failures had occurred in the 1840s, when the two ships of a British expedition commanded by Sir John Franklin were trapped in ice off King William Island for a year and a half; all hands ultimately died, many while trying to walk out, and some were apparently cannibalized by their starving shipmates. Such precedents hadn't worried Comer, because he had his seaplane with him. But neither had he expected to steam right through the passage in nineteen days, his path barely encumbered by ice. That unnerving experience had gotten him concerned about global warming—and, being the sort of man who creates $2-billion companies from scratch, he wanted to take action. As he began to educate himself on the subject, one name he kept hearing was that of Broecker, who as long ago as 1975 had been one of the first scientists to warn of the dangers of global warming. Would Broecker like to come out to Chicago for a meeting—soon, please?

As Broecker read this, he was sitting in his office at the Lamont-Doherty Earth Observatory, on the Hudson River about ten miles north of New York City. It was a room he had sat in for more than forty years, in a building he had helped build in 1954—a rambling, one-story cinderblock building, frequently added on to since. Morning light was streaming through the bank of windows on one wall, past the twenty-foot-long stuffed blue snake that hung above his expansive wooden desk. It was falling in dusty beams onto the bearded lady mannequin and the poster of Dolly Parton reclining on a haystack; onto the dozens of smaller photos of the graduate students and postdocs he had mentored; onto the long line of bound PhD theses that filled his bookshelf; and onto the motley assortment of curios and mementoes that a scavenging mind had collected over four decades of scientific roaming, on land and at sea. Later, after they had gotten to know each other, Comer would refer to Broecker's beloved Geochemistry Building as the "pigsty," but Broecker was happy there, like a pig in mud. Science is his life—and he had managed to make that office and Lamont into a world center of climate science.

Broecker wrote back that his teaching duties precluded a trip to Chicago for at least two weeks. A few days later he got a phone call: Comer couldn't wait that long, and would come to him. They met for breakfast at the Clinton Inn in Tenafly, New Jersey, not far from Broecker's home, and not far from the airport where Comer's jet had landed. Scientist and billionaire immediately hit it off. Both had been born within a few years of each other—Comer in 1927, Broecker in 1931—to working-class families in Chicago; Comer had never gone to college. Both men were straight talkers, both were at the top of their fields, and both abhorred red tape. They even liked the same breakfast, eggs sunny side up.

Since his trip through the Northwest Passage, Comer had sold Lands' End and been diagnosed with advanced prostate cancer. He wanted to do something about climate, and do it fast. Broecker, for his part, had been thinking of retiring. Comer's enthusiasm rejuvenated him. Over the next four years, Comer would invest his own money in climate research in a uniquely red-tapeless way: by giving it to established scientists to hire graduate or postdoctoral fellows. Broecker would select many of those mentors and give the program its scientific focus. Somewhere along the way Comer suggested Broecker write a popular book about climate. Broecker decided he needed the help of a popular science writer and invited Kunzig, whose book about oceanography he had liked, to collaborate with him. This book is the result.

Broecker has been studying climate, and in particular the way it has changed in the past, for more than half a century now. His career has coincided precisely with the emergence of man-made global warming as a problem. In the summer of 1955, while Broecker was a graduate student collecting some of the first radiocarbon dates of the end of the Ice Age, Charles David Keeling was making the first reliable measurements of the amount of carbon dioxide, or CO_2, in the atmosphere. Within a few years Keeling would report that the CO_2 concentration was rising—as it has continued to do relentlessly ever since, in direct relation to the use of fossil fuels. In recent years the rise has even accelerated, thanks largely to the coal-fueled economic booms in China and India, but also to the failure of the industrialized countries to restrain their own emissions.

The volume of research has followed the same upward trend; climate science is incomparably more sophisticated and specialized than when Broecker started out. Since 1990 the Intergovernmental Panel on Climate Change (IPCC), made up of hundreds of scientists from around the world, has tracked the swelling mountain of evidence in a way that no individual scientist possibly could. In its latest and most urgent report, released in 2007, the IPCC said the evidence showed unequivocally that Earth is warming; eleven of the last twelve years rank among the twelve warmest since 1850. The report concluded with more than 90 percent certainty that the warming is caused by greenhouse gas emissions—and thus that it will continue. Reviewing the forecasts from more than a dozen computer models, the IPCC gave as its "best estimate" that Earth's average temperature would warm anywhere from 1.8 to 4.0 degrees Celsius (3.2 to 7.2 degrees Fahrenheit) by the year 2100, depending in part on how much CO_2 we emit between now and then. The "reasons for concern," the panel said soberly, had gotten stronger since its last report in 2001.

To anyone paying attention to the news these past few years—of the rapidly shrinking Arctic ice cap, for instance, or of glaciers in Greenland and West Antarctica accelerating into the sea, or of the prolonged severe drought in places like Australia and the American Southwest—the IPCC report seemed sober to the point of understatement. The IPCC is a huge committee (which is one reason Broecker has always steered clear of it), and like most committees it is deeply conservative. That's why its reports have been so valuable—precisely because they are *not* alarmist, and if anything tend to stay behind the science rather than get ahead of it. In particular they do not do full justice to one of the biggest "reasons for concern"—what the IPCC calls "risks of large-scale singularities," and what Broecker once referred to, more colorfully but no less euphemistically, as "unpleasant surprises in the greenhouse."

That was in an article published in *Nature* in 1987. He was by then already deeply worried about the future of climate. Computers were not as fast and climate models not nearly as developed and well tested as they are now, but simple physics already made clear that temperatures would rise as atmospheric CO_2 did. And what troubled Broecker most was the lesson that he and other researchers were beginning to glean, not from computer forecasts but from cores drilled through the ice on Greenland and the sediments on the Atlantic sea floor—that is, from the solid records of climates past. The lesson has since been confirmed repeatedly, in studies that Broecker has devoted much of his time to analyzing and catalyzing: climate is not stable. On the contrary, it is a tetchy beast, subject to large and abrupt mood swings.

In Greenland at the tail end of the last ice age, temperatures rose to their present average—an increase of around 20 degrees

Celsius—in just a few decades, then stayed that way for millennia. Similar swings happened repeatedly during the Ice Age itself. Broecker's theory is that they were caused by a sudden jamming or restarting of what he dubbed the conveyor belt, a globe-spanning system of ocean currents that transports heat to the North Atlantic. The events of the Ice Age won't be repeated in the same way in a warming world—the conveyor is not likely to switch off in the next century—but there are likely to be other switches in the climate system that we understand much less.

So far we can only speculate what they are, and what flips them. What will happen to climate as a whole, for instance, when the Arctic ice cap that Comer skirted on *Turmoil* disappears altogether, as it looks set to do in the coming decades? How stable are the Greenland or West Antarctic ice sheets? If either one melts, sea level will rise between five and seven meters, and many of the world's great cities, including London, New York, and Shanghai, will be partially inundated, to say nothing of Kinshasa, Lagos, and most of Bangladesh. If that process happens over a millennium, as the IPCC still assumes, then the inhabitants will have time to adapt or move. If one of the ice sheets collapses in a century or two, then the "reason for concern" will be much, much stronger.

One of Broecker's biggest contributions to science has been this fundamental idea that climate does not always change smoothly, but that it also shifts abruptly between discrete states—that it has tipping points, to use the current buzzword. For a time, though he is by nature far from gloomy, that realization made him rather gloomy about the future. The massive emission of greenhouse gases seemed like a good way of pushing climate toward a series of tipping points, and there seemed no prospect for restraining those emissions. Indeed, they are still growing now, in spite of the Kyoto agreement, in spite of all the publicity that global warming has received lately, in spite of the IPCC having recently shared the Nobel Peace Prize with Al Gore. A quarter century after Broecker started thinking about abrupt climate change, solar panels and windmills have become familiar parts of the landscape in some places, but they contribute a tiny part of the world's energy supply. Consumption of fossil fuels is still growing.

That is not in itself a bad thing. Urgent as it is, global warming is not the most urgent problem for most of humanity; human misery is. Fossil fuels have lifted people in the industrialized countries out of misery, allowing the average person to live like a preindustrial king. The correlation between energy use and quality of life, as measured by such indicators as infant mortality, life expectancy, and literacy, breaks down past a certain threshold of development—western Europeans get along as well as Americans on half as much energy—but up to that threshold it is direct and strong. And that tells you right away that global fossil fuel use is going to increase, not decrease, in the decades ahead, no matter what course the industrial countries might take. China now uses around a quarter of the energy per capita that, say, Germany does; India uses around an eighth. Those two countries have well over a third of the world's population. As they reach for the quality of life found in the West, their use of fossil fuels will soar—because fossil fuels will remain, for the foreseeable future, the cheapest source of energy.

Burning fossil fuels is not bad; what is bad is dumping the waste into the atmosphere. There is a direct analogy to eating food, which is also not a bad thing. When we burn food in our bodies, we create waste too, and for centuries we simply dumped it wherever we liked. But as our numbers increased, and cesspools and privies got too close to wells, cities in America and Europe regularly endured not just foul smells but epidemics of typhoid fever and cholera. Today billions of people in poor countries still drink contaminated water; the World Health Organization estimates that six hundred thousand die of typhoid fever every year. The rich countries, however, have nearly eliminated such diseases, in part by building sewers and sewage treatment plants.

If we are to avoid dangerously warming the planet, we need to figure out how to build the equivalent of a sewage system for carbon dioxide—and what makes Broecker more hopeful these days is that the task now seems doable. In the final chapters of this book we present a vision for a carbon disposal system, a vision that draws heavily on the work of Klaus Lackner, a colleague of Broecker's at Columbia University. The technology for capturing carbon dioxide at the smokestack of a coal-fired power plant, before it even gets into the atmosphere, already exists, though it is still too expensive. The technology for "sequestering" carbon dioxide in deep rock formations is being tested at various sites around the world, and the results look promising.

The final and most novel element of Lackner's vision is just now taking shape, in prototype form, in a warehouse in Tucson, Arizona: it's a scrubber for removing carbon dioxide from ambient air. Such a machine is absolutely essential, because so much of our CO_2 comes from cars and planes. There is no prospect for a device that would capture such waste at the tailpipe or the jet engine, like the bags that are sometimes strapped to the rear ends of carriage horses; each vehicle produces too much CO_2. On the other hand, since carbon dioxide mixes quickly through the atmosphere, it doesn't matter where we take it out: if we take it out anywhere we benefit everybody. If we can find an economic way to scrub the atmosphere on a large scale—the underlying science is not complicated—we may one day be able not only to stop the rise of carbon dioxide in the atmosphere but perhaps even return it to the level we want.

It sounds like a utopian scheme; it sounds like too big a job. But cleaning up sewage is a big job too. A lot of the infrastructure for doing so, which we now take for granted, is more recent than young people might realize. In America, most sewage still flowed raw into rivers and the sea as late as the 1960s. But since the passage of the Clean Water Act in 1972, under President Nixon, the United States has invested more than $200 billion in sewage treatment. More than $100 billion of that came from the federal government. Even now, many of our waterways are not clean—but we have come a long way.

In the mid–nineteenth century, when the first municipal sewers were being built in America, there were plenty of sewage skeptics. For a while the science demonstrating the connection between sewage and disease remained uncertain; Pasteur and Koch were just then establishing the microbial theory of infectious disease. Even after the science was settled, however, and even after many thousands of people had died, some people still argued vehemently that the good old cesspools were good enough. But eventually the sewage skeptics faded away, a few no doubt from cholera and typhoid fever. People in the United States, as in other developed countries, came to accept that they had no fundamental right to dump their waste where they pleased, and that they should be willing to pay to dispose of it properly.

The same change in public opinion seems to be happening now with regard to carbon dioxide. The amount of carbon dioxide the United States emits each year, around 6 billion tons, is only around a tenth the amount of wastewater coursing through the country's sewers. Cleaning it up would be a big job—but not too big if we decide to do it.

Ultimately, of course, it would be better not to emit carbon dioxide at all. And ultimately solar and wind and fusion energy may make that possible—but we can't count on it happening, and on the fossil-fuel era ending, in this century. Yet the message from the study of past climates is that the time for stopping the increase in atmospheric CO_2 is now. Skeptics often find a strange solace in the knowledge that climate varies naturally, as if that somehow disproves the fact that we are rapidly changing it ourselves, or as if it somehow implies that climate change is inevitably benign. When you have explored the Ice Age for as long as Broecker has, and especially the wild swings that happened within the Ice Age, you don't think natural climate variability is benign.

Most researchers interested in the Ice Age have on their shelves a copy of *The Glacial World According to Wally*, one of the informal textbooks Broecker occasionally dashes off for his students, publishes himself, and sells at cost to whoever wants one. A few years ago he wrote one that has a cartoon on the cover. It shows a boy with glasses and unruly hair—he resembles Broecker a bit, but he is meant to be Everyman. The boy is poking a stick, labeled CO_2, at a Large-Scale Singularity, which is represented as a fire-breathing dragon rather than a long blue snake. The book is called *Fossil Fuel CO_2 and the Angry Climate Beast*.

FIXING CLIMATE

ONE

Pyramid Lake, 1955

The boy's journey to meet the beast began in Los Angeles, in the summer of 1955. Broecker arrived there at the end of August on a DC-7 from New York. It was his first transcontinental flight, and he spent it with a geomorphology book open on his lap, trying to understand the landscape from twenty thousand feet. Back then the pilots made a point of showing you the Grand Canyon at least. It was still a long, noisy flight, though—more than eight hours. When Broecker finally reached Los Angeles, he couldn't see much; the plane descended into LAX through thick smog.

Los Angeles was booming; its population had increased by half in the 1940s and was on its way to doing the same again in the 1950s. The city we know today, a megalopolis built around cars, was taking shape. The network of streetcar lines, which as recently as the gas-rationing days of World War II had carried more than 100 million riders per year on more than a thousand miles of track, was in its final agonies. During the war, people had mistaken some of the first severe smogs for Japanese gas attacks, and even by 1955 the true equation—cars plus sun equals smog—wasn't firmly anchored in the public mind. That summer workers were ripping up the rails of the Glendale-Burbank line, once one of the busiest. Congress had passed its first air pollution law in July, providing modest funds for research, but the catalytic converter was still twenty years away. Ozone levels in Los Angeles were higher that September than they've ever been since. City officials talked about shutting down refineries and gas stations; the governor talked about declaring the city a disaster area. The heat made things worse. On September 1, the temperature rose to 110 degrees Fahrenheit—an all-time record for Los Angeles. (It would not be equaled until 1988, the year global warming first dominated the headlines.) Around the city, hundreds of old people died from the heat that day.

That was the day Broecker delivered the first scientific talk of his life. The lecture hall was gorgeously situated, in the Southwest Museum of the American Indian—a Mission Revival–style building on a hill northeast of downtown, out toward Pasadena—but it was un-air-conditioned, and it was stifling. Through the tall, arched window at the back of the hall, Broecker would on a clear day have seen the Pacific, thirty miles away; but on September 1, 1955, he couldn't even see the Los Angeles skyline, five miles away. On either side of Broecker, the walls were lined with display cases of Indian artifacts. Before him sat eighty or so archaeologists, a mixed crowd of seasoned diggers and more scholarly university types. The diggers and the scholars, Broecker had already gathered, were in a spat about when Indians had first arrived in North America from Asia, the diggers favoring an early date. Broecker knew nothing of all this and little of public speaking. Painful memories of a college speech class came bubbling up, along with the question of just what he, a twenty-three-year-old graduate student in geochemistry, was doing there, delivering a special address to the 2nd Great Basin Archaeological Congress.

He was following orders, for one thing. His boss back at Lamont, Columbia professor J. Laurence Kulp, had been too busy to attend the congress himself. Like Broecker, Kulp wasn't much interested in archaeology. But the geochemical technique they were using at Lamont was of extreme interest to archaeologists—indeed it was revolutionizing the field. Radiocarbon dating had been invented just eight years earlier by Willard Libby at the University of Chicago, and there were still only a few labs in the world capable of doing it. Kulp had put Broecker in charge of the one at Lamont; he had even given the kid an assistant. Green as Broecker was, he knew all about radiocarbon dating. All he had to do was explain it to a bunch of archaeologists who knew next to nothing. That realization got him through the dreaded talk in the leaden heat. He may even have sounded a little smug.

Afterward, in any case, the man who had helped organize the conference strode up the center aisle and stood facing him. Phil Orr, archaeology curator at the Santa Barbara Museum, was clearly more of a digger than a scholar; although he smoked a pipe, it had a cigar butt in it. He was a short man with a potbelly stuffed into jeans and cowboy boots. His face was shaped like an interstate-highway shield—a wide forehead, uncluttered by hair, narrowing to a pointy, straggly bearded chin. That forehead overhung deep-set eyes that seemed made to squint. Orr eyeballed Broecker.

"Kid," he said, for he was plenty old enough to be Broecker's father, "I can see that you know a lot about physics and math. But I also see that you don't know a goddamned thing about the earth."

He paused to let that sink in, and to relight the cigar butt.

"Come with me for three weeks and I'll change your life."

Broecker's life back home contained outsize responsibilities. At twenty-three, he had a wife, Grace, and two daughters under the age of three. (Four more children would arrive before the decade was out.) They were all surviving on a graduate student's stipend, in the hope that the graduate student would soon make something of himself—get a PhD, for a start. Broecker called Grace. She said that as long as Orr paid for everything, Broecker should go along on this road trip or field trip or whatever it was.

Two days later the two men hit the road in Orr's old station wagon. They climbed out of the brown-aired valley and crossed the Mojave Desert, heading north on Highway 395. Following the Los Angeles Aqueduct to its headwaters, along the eastern flank of the Sierras, they drove up the Owens Valley, past Mono Lake, and on into Nevada. Along the way Broecker got to know Orr. The résumé was unusual. Orr had been raised on a Wyoming ranch, had trained as a vertebrate paleontologist, and had once stuffed a panda for Chicago's Field Museum. He was also a longtime member of E Clampus Vitus, the fraternal order of pranksters, Mason-mockers, and stuffed-shirt deflaters. This last item was particularly endearing to Broecker; the young man guffawed at the story of how Orr and a few other Clampers, dressed in tails and top hats, had driven a hearse to a Fourth of July picnic, solemnly marched a casket out into the midst of the festive gathering, and then opened it to reveal a case or two of beer on ice. By the time the two men pulled into

Carson City, where Orr had once mounted a coal mine exhibit for the local museum, they understood each other. Orr deposited Broecker at a motel, then disappeared—to go out drinking or to visit a girlfriend, Broecker was never sure which.

Nevada is empty now, but it was far emptier then; the population was around a tenth of the current one. Carson City, the capital, was a large village. When you left Carson City or even Reno, you really left—you did not pass miles and miles of subdivisions. Instead you plunged straight into sagebrush desert—into "savage nakedness," as John Muir once described it. The land was naked because it got around seven inches of rain a year. When Orr and Broecker were there, in late summer, the rabbitbrush scattered among the sage was blooming bright yellow, adding a little adornment, but the landscape was still basically brown. Around thirty-five miles northeast of Reno, the station wagon chugged up a low pass in the Virginia Mountains. From that crest, Orr and Broecker glimpsed the basin beyond. The sight startled Broecker just as it startles everyone the first time.

Nevada is famous for its basins and ranges; they march in orderly succession across the state, like waves breaking on the Sierra Nevada. But this basin was different: it was filled with blue water. John C. Frémont, the first white man to be startled by the sight of a twenty-five-mile-long turquoise lake ringed by brown desert mountains, called it Pyramid Lake, because it had a small pyramid-shape island off its eastern shore. To Frémont, in 1844, the lake would have been far more inexplicable than it was to Broecker. The notion of a worldwide Ice Age was only just beginning to be toyed with then by European geologists, and Frémont didn't have Phil Orr with him to explain what the Ice Age had done to western Nevada.

Pyramid Lake, Orr told Broecker, is a remnant of the Ice Age, and of a mighty ancestor called Lake Lahontan. The ice sheet that covered Canada and much of the Midwest never made it to this part of Nevada; nor did the mountain glaciers that surged out from the Rockies and the Sierras. But the glacial climate seems to have been colder in the Great Basin, and it was certainly much wetter. With no outlet to the sea, and without today's desert heat to evaporate it, all that water falling from the sky and rushing down off the Sierran glaciers just piled up. At its highstand, Lake Lahontan was larger than Lake Ontario; it covered much of western Nevada, occupying seven distinct basins that formed a ragged necklace around the Trinity Range. As the climate warmed and dried, the lake sank below the sills that connected the basins, one after another. The basins that weren't fed directly by rivers from the Sierra dried up. The only ones left today are Walker Lake, terminus of the Walker River, and Pyramid Lake, terminus of the Truckee River and also the deepest part of Lake Lahontan.

As Lake Lahontan rose and fell, its waves cut terraces in the mountainsides wherever it paused for a while. Sometimes it even cut caves. By 1955, Orr had already dug extensively in a place called Fishbone Cave, high above Winnemucca Lake in the next basin east of Pyramid. He had found human artifacts and bones, which he believed were very old—he belonged to the camp that claimed the ancestors of today's Indians had gotten to North America early, even before the end of the Ice Age. But he needed radiocarbon dates to test that hunch and to wave at his scholarly foes, and radiocarbon dates were expensive and hard to come by. Orr thus had a sharp and specific interest in the Eastern greenhorn sitting in his passenger seat. He wasn't teaching Broecker Pleistocene geology and Western outdoor savvy purely out of the goodness of his heart. (Those trails through the sagebrush, Orr explained to his gullible young friend, had been cut by an extinct species of cow, well-adapted to steep slopes because its legs were shorter on one side.) Over the next few years Orr got around fifty radiocarbon dates for free.

But in return, as promised, he changed Broecker's life. The two men spent several days bouncing around Pyramid and Winnemucca in the station wagon, which sucked in dust and covered them at the rate of about a quarter inch a day, and which periodically also got stuck in the sand. They scrambled up hillsides and crawled into caves, hammering bits of tufa off ancient lake terraces and shoving them into bags. Tufa is a weird kind of calcium carbonate that forms when calcium from lake-bottom springs reacts with carbonate dissolved in the water; sometimes algae may help the reaction along. At Pyramid Lake tufa comes in giant towers that look from a distance like a city skyline, in giant branching bulbs like broccoli, and in shattered slabs. What mattered to Broecker was only that tufa contained carbon. Back in his radiocarbon lab, he would be able to find out when each tufa sample had precipated out of the lake—and if it had precipitated on an ancient shore, as a lot of it had, Broecker would know when the lake had occupied that shore. It was the highest shore he was most interested in, to find out just when the lake had last been at its largest. The highest shore was called Lahontan Beach.

On Lahontan Beach, Broecker discovered climate and got a glimmer of what a fascinating beast it could be. At a place called Marble Bluff, which overlooks the mouth of the Truckee at the south end of Pyramid Lake, the cobbles once lapped by the ancient lake are visible near the top of the steep hill. From there, your car parked on the road below is a dot, and the lake is another mile and a half beyond it. The water is more than five hundred vertical feet below you. That's how much water has disappeared since Lahontan times. Even here, the Ice Age—which cooled the planet by fewer than 5 degrees Celsius—caused a big change in the landscape.

If you look off to your right from Marble Bluff, to the east of Pyramid Lake, you see a low rise called Mud Sill, and beyond it the next basin, flat and glimmering white in the sun. That's Winnemucca Lake. It is no longer a lake but a playa—a salt flat. But unlike the playas north of Pyramid—including the hundred-mile-long Black Rock Desert, a Lahontan lake bottom where people now set land speed records—Winnemucca Lake didn't dry up thousands of years ago. It dried up in 1938. The cause was a modest little structure called the Derby Dam, twenty miles east of Reno on the Truckee. The Derby Dam was the first dam completed by what is now the U.S. Bureau of Reclamation, in 1905. Its purpose was to divert Truckee water into a canal that would carry it thirty-two miles southeast into the Lahontan Valley, so that farmers there could make the desert bloom with alfalfa, which could then be fed to cows. After the dam was finished, Pyramid Lake dropped steadily until finally it no longer overflowed Mud Sill. After that Winnemucca was doomed. It had never been a deep lake—some called it Mud Lake, because in many years it was more like a marsh—but it had been a haven for migratory birds. Drying up Winnemucca was an unintended consequence of a well-intentioned act, one that many people in Nevada, although not the Lahontan farmers themselves, now consider a mistake.

The environment changes on its own, and man changes it too—locally, as at Winnemucca, but also globally, as the ensuing decades would make clear. Small changes can have big effects. Broecker's brain was not primed then to plunge into such murky waters. He was a fresh-faced, clean-cut young man with the devil in his eye. He hadn't worried about the smog and the traffic jams back in Los Angeles. He had driven by Owens Lake, which Los Angeles had dried up by channeling the Owens River into the Los Angeles Aqueduct, and which had now become a serious dust hazard—it had not made an impression on him. And he had driven and walked for days around Winnemucca without even knowing that its demise had been man-made. That's the way things were in 1955; the environment was only just starting to register on most people's minds. Broecker was hardly unusual.

The problem of understanding Earth's changing climate, as it revealed itself to Broecker then for the first time, was not yet a gloomy problem. It was not about saving the world. It was just a giant puzzle, and puzzles were what Broecker's brain was good at. Puzzles were fun. At Pyramid Lake, home, before the water level dropped, of forty-pound cutthroat trout, Broecker was hooked for life.

TWO

Finding Science

Though he was not aware of it at the time, Broecker had spent the first eighteen years of his life on the bottom of an Ice Age lake. When Lake Lahontan was at its last highstand—11,500 years ago, according to Broecker's initial radiocarbon date, but later work pushed that date back to around 15,000 years—Oak Park, Illinois, was still covered by ice. It was the last gasp of the last Ice Age, the Wisconsinan period, and the ice sheet, retreating north from central Illinois, had paused for a spell. As ice laden with rock and dirt continued to flow into the snout of the glacier and melt there, it built up a large moraine. When the glacier started retreating again, withdrawing into the Lake Michigan basin, the moraine acted as a dam. A widening arc of meltwater pooled in the space between the moraine and the receding glacier, forming a lake that geologists call Lake Chicago. All of what is now the city of Chicago was submerged under it, and most of what is now Oak Park, which is eight miles due west of the Loop.

Oak Park is flat, flat, flat—a fact that young Broecker would learn to appreciate as he tooled around on his bicycle delivering newspapers. But if you go to Lake Street in the center of the village and look into Scoville Park, you will see that the grass slopes gently up to the northwest, toward the war memorial. Or if you drive west on, say, Berkshire Street, and pay close attention, you will notice at one point that you are crossing a low rise. In both places you will have encountered the Oak Park Spit—a long, low bar of sand and gravel now hidden by grass and asphalt. Currents running down the west shore of Lake Chicago deposited that sand; the bar jutted out across the mouth of an inlet called Des Plaines Bay.

Once the ice and then Lake Chicago had completely retreated, becoming Lake Michigan, that bay became the valley of the Des Plaines River, which flows through River Forest, the town just west of Oak Park.

Meanwhile the lowly Oak Park Spit, fifteen or twenty feet high at most, became a continental divide. Waters west of the spit flowed into the Des Plaines River, then into the Illinois, the Mississippi, and the Gulf of Mexico; waters east of the spit flowed into the Chicago River, Lake Michigan, the St. Lawrence, and the Atlantic Ocean. That pattern endured for ten thousand years or more—until 1900, when the young and booming city of Chicago, concerned about all the sewage its river was dumping into its water source, Lake Michigan, decided to dam the river and excavate a channel through the low divide. That reversed the flow of the Chicago River and flushed the sewage down the Mississippi. It was widely considered one of the engineering wonders of the age, except in St. Louis and other towns along the Mississippi. A few years ago the Oak Park Rotary Club went around the village and put up signs on the spit that read "Continental Divide: Historical Boundary." Since 1900 all of Oak Park's runoff has ignored that boundary and run to the Gulf of Mexico.

When the first settlers arrived in Oak Park in the 1830s, most of the place was still swampy, and so they settled on the high, sandy spit, near what is now Lake Street. By the time Broecker's parents arrived in the 1920s, Oak Park had a fast connection to Chicago via the Lake Street Elevated Railroad and was well-established as a leafy haven for people fleeing the city. Most of those people were of German descent, such as Wallace Charles Broecker—his grandfather had emigrated from Germany—or of English descent, such as Edith Smith Broecker. Oak Parkers were a God-fearing lot: by the 1930s the village had thirty-eight churches and no bars. It had been dry since 1872. Yet, something about the place fostered creativity. Among the people who lived in Oak Park in the first three decades of the century were Frank Lloyd Wright; Ernest Hemingway; Edgar Rice Burroughs, author of the Tarzan books; Richard Sears, founder of the Sears, Roebuck Company; Ray Kroc, founder of McDonald's; and James Dewar, inventor of Hostess Twinkies.

The house Wallace and Edith Broecker bought in 1920, about

half a mile from the Ice Age beach and the center of Oak Park, on South Scoville Avenue, was not designed by Frank Lloyd Wright. It had a front porch, three rooms, and a kitchen downstairs, and two small bedrooms upstairs under the eaves. The stairs were in the back outside, which was something you noticed in winter. There was one bathroom, downstairs. The Broeckers had both grown up in South Side tenements near the Union Stock Yard; it was the period when Carl Sandburg was celebrating Chicago as the "hog butcher for the world" and as a "tall bold slugger set vivid against the little soft cities." To the young couple, the modest little house on a wide, shady street must have looked pretty good.

But by the time they put five kids of their own into those two upstairs bedrooms, it would come to feel a bit close. Young Wally was the second of the five. He was born in 1931 (the same year as Twinkies). That year nearly half the people in Oak Park defaulted on their property taxes; the Depression had hit the village hard. Thousands of men in the town of sixty-four thousand were unemployed. The Red Cross was distributing food. The Broeckers never needed that kind of help; Pop Broecker owned a Standard Oil gas station on Chicago's North Side, on the corner of Chicago Avenue and Larrabee. It was a rough neighborhood, but right across the street was a large Montgomery Ward's warehouse that kept Broecker in business, parking and servicing the cars of the executives.

He worked long hours—out of the house by 5 a.m. every morning, not home again until 7:30 p.m., even on Saturdays. In the evenings he would fall asleep on the couch reading the *Tribune*. He was a handsome man, with brown hair and regular, angular features, and always carefully dressed; a hardworking but cheerful man who was always whistling—that's how the kids would know he was finally home. He was gregarious too, capable of going into a diner for a cup of coffee and not coming out for two hours. But he didn't talk much with his kids about anything important. In moments of family crisis he would sometimes hide behind the *Tribune*; the gas station in a slum was to him a simpler place. Wally's older sister, Jeanne, who had a rebellious streak, once said of her father that if he were to find his entire family stretched out dead, "he would whistle through the mortuary." It was unfair perhaps but not entirely untrue.

Home life with Edith was not simple. Tall and good-looking like her husband, she was more intellectual—like him she hadn't been to college, but she was an avid reader who listened to classical music on the radio and even wrote poetry. For most of her adult life, however, she suffered from recurring, severe depressions that would send her to her bedroom and sometimes to a hospital for weeks. The children would then be farmed out to other families, or a friend named Mrs. Austin would come to stay with them. Edith Broecker was a devoted wife and mother when she was there, but many times she was not there.

She was an atheist when she married, and she declined to convert to her husband's Catholicism. (Wally never did meet his paternal grandparents.) But when she and her husband were later "saved" at the Harrison Street Bible Church, on the southern edge of Oak Park, Edith threw herself into her new religion with enthusiasm. She would listen intently to the sermons and jot down, on the flyleaf of her Bible, sentences that moved her. The famous line from St. Augustine, for instance, confessing a youth of wandering far from God: "And I became to myself a wasteland." All her life Edith was a learner and a seeker.

Inheritance sometimes seems to operate in simple ways. Wally Broecker got his father's disposition and his mother's mind. Edith Broecker got a son who was a learner too—but a heretic who would later doggedly try to convince her, on scientific grounds, that the Flood couldn't have happened. And who would, regardless of Genesis and its interpreters, insist that Earth couldn't be younger than those tufas he had radiocarbon-dated at Pyramid Lake.

People who in later years crossed swords with Broecker in the arenas of science might be satisfied to learn that his family called him the Little Kaiser, and that before that he was an extremely fussy baby. He was apparently allergic to cow's milk, a relatively common affliction that usually subsides by the time the child is two. It can be a long two years for the mother, however, and toward the end of it Edith was at the edge of despair. One day, perhaps while walking her son in his pram, she encountered a woman whom the family would come to know as Dr. Hess—a strange woman from the Ozarks who would change all their lives forever.

Dr. Hess practiced naprapathy, an offshoot of chiropractic that originated in Illinois around the turn of the century and is still practiced primarily in that state. (Three other states now offer licenses.) Naprapaths treat pain by manipulating ligaments, tendons, and so on, on the theory that the origin of the pain is some kind of imbalance in the connective tissue. Dr. Hess quickly decided why young Wally was always squalling. "His legs are out," she explained to his parents—meaning that one leg was longer than the other. Her treatment consisted of laying the boy down and cranking on the short leg and shoving on the long one until she was satisfied they were aligned. To the sleepless parents, her very first efforts seemed to bring peace to their child at last.

For years after that Dr. Hess returned regularly to the Broecker household to give tune-ups, not just to Wally but to the whole family, lining them up like logs on the living room floor. She became deeply involved too in treating Edith's depressions. Wally's younger brother, Howard, eight years his junior, remembers being taken to Dr. Hess's basement office on Ridgeland Avenue in the late 1940s to be connected to Edith by means of electrical pads and cables, which Dr. Hess thought might conduct some beneficial energy from the little boy to his mother. There was more than a bit of the quack in Dr. Hess. Her main influence over the Broecker family, however, was not medical. It was religious.

Dr. Hess belonged to the Harrison Street Bible Church. After her apparently successful manipulation of Wally, his grateful parents accepted her invitation to try out the church. It was a small congregation, only 150 souls or so, but tight-knit and energetic. The pastor, T. Leonard Lewis, held revival meetings in a tent in a vacant lot adjacent to the church. After many years of erring in the wasteland, Edith and Wallace Broecker were revived at Harrison Street—and their conversion was all the more powerful for coming relatively late, when they were already in their thirties. At Harrison Street they found not only faith but also a community of friends that would rally around them in time of need.

Henceforth religion dominated the Broeckers' lives. The only day the children saw much of their father was Sunday, and the family spent most of Sunday at church. There was Bible school at nine-thirty, followed by Morning Worship at eleven. They went home for dinner then, sometimes bringing the choir director with them. Then at six the children returned for Juniors' Meeting, and at eight there was gospel service. There was a prayer meeting on Wednesday evenings too, which Wally Broecker's brother and sisters attended, but from an early age he found a way to get out of it. It was just too much.

And of course, being a good evangelical Christian is not simply a matter of putting in the hours in church. The faithful at Harrison Street subscribed to a literal interpretation of the Bible, but like many fundamentalists they also adhered to many rules that are not mentioned in the Book. They didn't drink or smoke, play cards or dance ("a vertical manifestation of a horizontal desire"), or go with boys or girls who did any of those things. Even tougher, they didn't go to movies—unless the church itself was showing *Thief in the Night* (a dramatization of the Second Coming and the Rapture) or some other doctrinally worthy film. Howard Broecker still remembers the first time he snuck out to violate the no-movie-if-it's-fun rule: he saw *20,000 Leagues Under the Sea*. As he sat in the darkened theater tasting the forbidden pleasure, he thought lightning was going to strike.

Wally Broecker snuck out to the movies too, but the other restrictions didn't bother him much. For one thing, he made friends at church or in the neighborhood who were subject to the same rules. By the time he became a teenager, the Reverend Billy Graham, recently graduated from nearby Wheaton College, was organizing Youth for Christ, an outfit whose motto was "Anchored to the Rock, Geared to the Times." That gave a Christian kid something to do on Saturday nights: he could go to a Youth for Christ rally.

But from T. Leonard Lewis to Edith Broecker to Billy Graham, no one ever quite managed to anchor Wally Broecker to the Rock. Even as a boy he found it hard to believe that the sun could have stood still for Joshua or that the Red Sea could have parted for Moses. All the hellfire talk he heard at Harrison Street, and to some extent even at home, sat uneasily with him. The story of *Thief in the Night*—Jesus' warning on the Mount of Olives that you better act right because you never know when He might return to destroy all sinners—was always less inspiring to Broecker than the story of the Thief on the Cross—the guy who lived a life of sin but then got Jesus to save him at the last minute. Not that young Broecker thought deeply about theological matters; he did not think deeply about any matters. And not that he was drawn to a life of sin per se. It was really only fun that he was after, and he never let fundamentalism get in the way.

He sailed through school. He was always good at it, especially math, and mostly indifferent to it. At home he would rip off his homework while listening to the *Lone Ranger* or *Tom Mix* on the radio, then beat it outside—or if the weather was bad, upstairs to his little room to build model planes. He hung them on threads from the ceiling; it was the era of the Battle of Britain and Pearl Harbor. But he much preferred to be outside in the back alley, playing basketball with his pals. Sports were the most important thing to Broecker in those days. He certainly did not waste his free time reading books. Reading was the one thing that was never easy; much later, when some of his own kids developed the same problem, he realized that he and they were probably dyslexic. It may have made him more sociable than he would otherwise be: early on he developed the habit of absorbing information by listening to people.

Edith watched his report cards closely, encouraged him, and seldom had cause to reproach him, for they were mostly filled with Excellents and Very Goods—except on the line marked "Comportment." There the grade would often dip. Edith would then keep the boy inside the house after school, one week for each notch below Satisfactory. It was young Wally's personal taste of hellfire; he loathed and feared it. He learned the value of not getting caught, and also of bending rules rather than breaking them with a loud snap. Being cooped up in the small house, especially after it got crowded with infants and toddlers—his siblings Judy, Howard, and Bonnie were born between 1938 and 1947—was not something an adolescent boy could be expected to relish.

From an early age he found ways to escape. He had a strong personality and always tended to seize whatever freedom was on offer; as long as he didn't get into obvious trouble, his parents—because of the other pressures they were under, or just because parents were less protective in those days—allowed him a long leash. Beginning in World War II, when paperboys were in great demand, he also had an independent source of income. Pop would wake him at 5 a.m., just before leaving for the gas station, and he would ride his bike down to the *Tribune* depot at Lombard Avenue, where he would spend half an hour or so folding and rolling the newspapers tight enough to be flung onto a porch or even a third-floor balcony. He kept the money he made in the soup tureen in the dining room. Every now and then he would find an IOU from Edith in there. Pop was sometimes tight with the household money.

Some of the money Wally made he spent on the Harrison Street el. For a nickel you could ride into Chicago. He and a friend would go downtown and climb the fire stairs onto the roofs of the tall buildings. Mom was at home, Pop was at the gas station, and unbeknownst to either of them, young Wally was there, on top of the world. He and his friend would make paper airplanes and launch them into the void. Sometimes too they would change trains downtown and ride the South Side el; the trestles were so narrow on that line that when you leaned out the windows, you could look straight down at the street. They would lean out the windows and try to drop marbles on the cars and people below. Broecker doesn't recall ever hitting anybody.

He doesn't remember much about the Museum of Science and Industry either, though he went there more than once on his illicit outings; it was right at the end of the South Side el. Often when an eminent scientist starts fleshing out the earliest entries of his curriculum vitae, it will include a scene in which the young whiz spied his first dinosaur skeleton, or the craters of the moon, or whatever, and his passion for science was thereby ignited. There are no such scenes in Broecker's childhood; they came much later. He didn't give a hoot about science then.

But there are scenes that show how much he liked to make things, and to make things work. Pop Broecker had a movie camera, and Howard Broecker has a few of the old snatches of film, transcribed onto videocassette. In one of the silent, jittery sequences, there is a quick glimpse of Wally. He is in his early teens, and he is driving down a shady street, smiling and swerving back and forth as if to show off the steering system of his strange-looking vehicle. It is a small wooden car that Broecker had designed and built himself—a Soap Box Derby car, except that it was motorized. In the local penny-saver Broecker had spotted an ad for a used Briggs & Stratton gasoline engine, rescued from some washing machine, and he had bought it with his paper-route money. In the first version of the car, the engine was mounted on a trailer that pushed the car along. Broecker could cruise Oak Park's flat streets for three hours before he had to pull into a gas station; he got a big charge out of yelling "Fill her up!" to the startled attendant. At twenty-five cents a gallon, that was around five cents. Sometimes the cops would pull him over, but as soon as they left, he would ignore their stern warnings and take off again. On the boulevards, with only his shoes as brakes, he could hit 30 mph—he knew because every so often he would pull even with a car and shout, "How fast am I going?" to the driver. He did not wear a helmet.

And so Wally Broecker's youth passed, lightly and happily. Those years included the Depression, World War II with its attendant rationing (though Pop had all the gas he wanted), a troubled mother, and frequent reminders of the potential for eternal damnation, but Broecker remembers them as happy. The year the war ended he entered Oak Park River Forest High School, Hemingway's alma mater. The pictures in *Tabula*, the high school yearbook, show a short, skinny, baby-faced kid—too short and skinny, to his great chagrin, for freshman basketball, so he managed the team instead, along with a good friend named Ernest Sandeen. By their senior year they were still standing next to each other in the yearbook, only now they were both wearing suits and loud ties; they had both made the four-year honor roll. Broecker had had a growth spurt by then, but he still looked young for his age. He was a late bloomer. His picture now hangs near Hemingway's in a kind of alumni hall of fame the Oak Park students set up in the 1980s; he and Papa are the only two who aren't wearing ties.

In their senior year Ernie Sandeen, who went on to become a religious historian and to write an influential history of fundamentalism, was president of the Chemistry Club. Wally Broecker, who went on to become a geochemist and a member of the National Academy of Sciences, did not belong to the Chemistry Club; his parents had given him a chemistry set once and he had hated it. Nor did he belong to the Biology Club, to the Radio Club, or to Newton, the physics club. What he remembers best about high school was shop—the wood turning and the metalwork. The Oak Park shop was well equipped; it had a forge, and each student had his own lathe. Wally made his mother a cherrywood salad bowl.

Harrison Street rules kept him somewhat isolated socially, but the summer after his junior year, at the Winona Lake Bible Conference near Warsaw, Indiana, he discovered girls—specifically, a cute little brunette named Grace Carder. Wally had not gone to Winona Lake for the preaching anyway. He had gone there with Sandeen to get away from the family and to make money working as a busboy in the cafeteria by the lake. Grace was a waitress next door at an ice-cream parlor called the Eskimo Inn. (Today they have six children, seven grandchildren, and five great-grandchildren.) Grace's parents were missionaries in the Canary Islands, and she attended a boarding school in Wheaton that was affiliated with Wheaton College. After that summer Wally started borrowing his father's Cadillac—money may have been tight, but Pop always had a nice car—to zip out to Wheaton to court Grace.

Thus when Ernie Sandeen announced after that summer that he was applying to Wheaton College, the idea of going with him made sense to Broecker on several levels—though he'd given little thought to college himself. He had worked in Pop's gas station, cleaning out a sordid, rat-infested basement that extended out under Chicago Avenue; he had even run the place once while his parents took a vacation up in Michigan. He had no idea what he wanted in life, but he knew that pumping gas wasn't it. His parents liked the idea of him going to Wheaton too. The pastor at Harrison Street, not to mention Billy Graham, was a recent graduate. It had a reputation as the Harvard of evangelical colleges, an ambitious but also a godly place, where students signed a pledge to adhere to the same sort of rules that applied at Harrison Street. (The pledge was modified in 2003 to allow students to dance under supervision, but not to drink, smoke, or fornicate.) Then as now, Wheaton was "committed to the principle that truth is revealed by God through Christ, 'in Whom are hid all the treasures of wisdom and knowledge.' " The motto "For Christ and His Kingdom" was inscribed in stone at the campus entrance.

It was a serious place, in other words—and it seemed to force into full flower the most profoundly unserious aspect of Broecker's character, one that can still be startling today to the uninitiated. He started pulling pranks at a young age and has never quite outgrown the habit. One of his earliest triumphs had come in fifth grade, when on a bitter winter morning he convinced many of the boys in Mrs. Ferguson's class at the Hawthorne School that what they ought to do during recess was go to the boy's bathroom and urinate in unison on the large cast-iron radiator. The smell, Wally predicted, without any particular knowledge of the chemistry, would force the school to close. As a result of some other minor offense, Broecker was locked in a coat closet at the time his friends executed his plan. He thus escaped the reprisals they suffered.

At Wheaton the pranks got more elaborate. In his junior year he was business manager of the yearbook, which position of influence allowed him to sign excuses that got him and his friends out of daily chapel. It also got him a key to the attic of Blanchard Hall, where college memorabilia were stored. That October, Broecker discovered an unused door leading from the attic into the bell tower, which was off-limits to students; the custodian kept the main entrance carefully padlocked. At midnight on Halloween night Broecker, Sandeen, and their roommates woke the campus with a loud tolling. When the custodian came racing into the bell tower, they exited through the attic and locked him in. He was forced to ring the bells again to summon the police. By the time the law arrived the Broecker gang had retreated to a ground-floor classroom. Broecker remembers vividly the flashlight beams coming through the windows and playing along the walls as he and his friends hugged the floor, out of sight.

There were other pranks that same year—a box of mothballs dumped on an unpleasant librarian by means of a rope-and-pulley system that allowed the dumpers to remain anonymous; a granite bench ripped from its prominent position on campus and buried. On that last job Broecker had the assistance of a friend named John Nuckolls, a rocketry enthusiast who had by then already called attention to himself at Wheaton with a serious accidental explosion; he would go on to become Edward Teller's right-hand man in the development of the H-bomb, and the director of the Lawrence Livermore laboratory. It was Broecker, however, who was hauled into the dean's office in the missing-bench case and threatened with expulsion. "You'll look pretty foolish," Broecker shot back, "expelling a student with one of the highest grade point averages in the college." He was by then beginning to question authority more vigorously. And in truth, although he enjoyed his time at Wheaton and got a solid education, and not just in the Old and New Testaments, he had become a bit of an outsider there.

The change had begun his sophomore year. In addition to going to chapel every day, Wheaton students were required to spend a week every year rededicating themselves to Christ under the guidance of a visiting preacher. That year the event took an extraordinary turn—it became a mass public confession. For three days and nights students lined up in the choir loft and behind the pulpit, waiting for their chance to proclaim their sins. The pressure to participate was intense. Broecker sat there in turmoil, brooding more seriously than he had ever brooded about anything. He certainly had sins—violations of the Wheaton pledge, for instance, that went beyond dancing. (He has never liked dancing.) But he knew some of the people who were confessing, and he knew they weren't being honest. They were holding back on the juicy stuff. The hypocrisy of the whole spectacle revolted him—people were pretending to believe in rules they couldn't really live by, and then pretending to confess their violations of those rules. Hypocrisy and dishonesty, Broecker realized, were the sins he could least abide. That day in chapel, he slipped the fragile line that had tied him to the Rock.

Billy Graham came to visit that same year, passing the assembled students in review like a general inspecting troops; Broecker was one of the lucky ones to be touched on the head. "Young man, the Lord is with you," Graham said. "You will do well." It was a risky prophecy. Broecker had never been a directed young man, but now he was truly adrift—cut loose from the strong framework his parents had built after that first chance encounter with Dr. Hess. Unbeknownst to him, however, he had already had his own life-changing chance encounter. Unbeknownst to him, there was a small conduit that led, like a wormhole into a parallel universe, from Wheaton College and the world of "truth revealed by God through Christ," to the Lamont Geological Observatory and the world of science.

At the near end of the conduit was a Wheaton student named Paul Gast; at the far end was a Wheaton alumnus named J. Laurence Kulp. Wheaton had the enlightened practice of assigning a sophomore adviser to each entering freshman, and Broecker had drawn Gast as his "big brother." Later it would become plain that Gast was on a trajectory similar to Broecker's—he came from a devout family of German origin, and he was on his way to becoming the head of NASA's moon rock program. (After he died in 1973, a ridge on the moon was named after him.) But at the time he was everything Broecker was not: responsible, studious, and serious. He even took his responsibility to Broecker seriously. Though they did not move in the same social circles, Gast not being the kind to assault unsuspecting librarians, when their paths did happen to cross, Gast would often stop to ask Broecker just what he intended to do when he grew up. Broecker would fob him off.

Meanwhile the Lamont observatory was also growing up. It had been born at the end of 1948, when Florence Lamont, widow of a wealthy financier, donated the family estate on the Palisades to Columbia University, and Columbia handed it over to geophysicist Maurice Ewing. Soon Ewing was storing ocean-sediment cores in the dining room of the old mansion, and the seismologists he hired were installing their instruments in the empty swimming pool and in the root cellar. Ewing hired Kulp to set up a geochemistry lab in the kitchen, where gas and water ran freely. Kulp had promptly gone to Chicago for several months to apprentice with Willard Libby, who had just invented radiocarbon dating. Kulp had also gone to Chicago to recruit students at Wheaton, his alma mater. One of his recruits was Gast, who spent the summer of 1951 working at Lamont, and who was hired to come back the next summer, after he had graduated.

In the fall of 1951, around the time Broecker was imprisoning the Blanchard Hall custodian, he ran into Gast again. Broecker was majoring in physics by then, because it was easy and he sort of liked it, but he had no notion of science—or of anything else—as a career. He did not even know what graduate school was. He had to appease Gast, however, and fortunately he had recently read a magazine article about actuaries. Their conversation went something like this:

> Gast: "Wally, what are you going to do when you get out of here? Now you're a junior—you've only got a little over a year."
>
> Broecker (brightly): "I'm going to be an actuary!"
>
> Gast: "Oh, come on, that's really boring. You don't really want to do that!"
>
> Broecker (testily): "Well, you got any better suggestions?"

Gast did, in fact: he could arrange a summer job in Kulp's lab. And so on June 15, 1952, Broecker and Grace—they had just gotten married, and she was pregnant—drove up the winding drive to the Lamont mansion. They did not stay in the mansion; Grace set up their first household in a hut at nearby Camp Shanks, from which troops had shipped out to Europe during the war. But Wally spent the summer working in the Lamont kitchen and especially in the basement of the mansion, where Kulp had installed his radiocarbon operation. It included mass spectrometers, Geiger counters, vacuum chambers for melting rocks, and a twelve-ton lead-and-iron box to protect the measurements from ambient radioactivity. Broecker had in his bags a "flip-flop circuit" that he had laboriously put together in his physics lab that year at Wheaton. It was not unrelated to the hardware he now had before him, but it was the extent of his experience.

He caught on quickly though. The Kulp lab had been radiocarbon-dating successfully since 1950, but by 1952 things had started to fall apart. Blank samples, which should have had no radiocarbon in them, kept generating lots of action in the radioactivity counter. One nagging problem was arcing between the central, high-voltage wire of the counter and the Pyrex tube that was supposed to insulate it; each electric discharge would spuriously mimic a radioactive decay. Broecker suggested replacing the tube with some Teflon insulation he found lying around the lab. It worked, and Kulp began to realize, This kid is thinking. Broecker began to realize, this kind of thinking is fun. But even after that fix, the counters kept overcounting—erratically, on some days but not others, such that no measurement could be trusted. The lab was paralyzed. Finally the new kid from Wheaton ventured another suggestion: could it be that they were detecting drifting radioactive fallout from the atmospheric nuclear tests that had recently begun in Nevada? That realization led to an entirely new research program, which was to occupy much of Kulp's time for the rest of the decade.

By the end of that first summer Broecker was practically running the radiocarbon lab for his busy boss. When the day came for him and Grace to leave for Wheaton, Kulp announced that for Broecker's own good, he could not be allowed to return to a college that had nothing left to offer him. Within two days Kulp had arranged for him to transfer to Columbia as a senior and to keep doing what he was doing for Kulp. Wally and Grace were not inclined to turn him down. They were expecting their first child within a few months. There wasn't really any going back to where they'd come from. And they never did.

In the end, if you're lucky, it is all about joy and where you find it. Edith Broecker found it in Scripture, in the little bits of revelation that helped her make sense of her life and that connected her to other people in a community of the faithful. Wally Broecker had passed into a different community with different joys, which were more suited to his constitution. Some were just variations on a kind of fun he had known before. Blowing bubbles of acetylene into the stairwell of the Lamont mansion, then poking them with lighted matches to create rising balls of fire—he had shown the same inclination and the same kind of skill at Wheaton. But in that summer of 1952 he also discovered a new and deeper kind of joy. Upstairs on the second floor of the mansion, in a converted bedroom and bathroom, a couple of scientists named Bruce Heezen and Marie Tharp were putting together a new kind of map of the seafloor—one that would show there was a volcanic ridge running down the middle of the world ocean, and that would help pave the way for the scientific revolution called plate tectonics. Down the hall, meanwhile, Ewing and his colleagues were starting to show that sediment cores from the ocean floor contained a record of Earth's past climate. And down in the basement, in the radiocarbon lab—well, to be a radiocarbon dater in the 1950s was to be like Galileo with his new telescope in the 1610s. Everywhere you looked you saw something no one had seen before. Discovery was easy and intoxicating. Within minutes you could know, say, when Cro-Magnon men had inhabited some cave in France, or when Lake Lahontan had lapped its highest shore. Sometimes in those early days Broecker would just sit there in the basement, watching the radioactive decays register on his counter, one by one.

Larry Kulp had remained a religious man—he had converted to fundamentalism as a teenager, rejecting his parents' milder Episcopalianism—and he worked hard to reconcile his two worlds. He wrote papers trying to convince his fellow believers that Earth could not possibly be only ten thousand years old. For a few years after Broecker arrived at Lamont, he attended Kulp's church, the Plymouth Brethren, out of respect for Kulp, but eventually he gave it up. A few of the Brothers came to his house to talk with him, to no avail; even his father-in-law, the missionary, home from the Canary Islands, could not persuade him. Broecker never quite swung over to a hard scientific atheism—he never became comfortable with the notion that the universe arose spontaneously from a vacuum, to which it may one day return. He just lost what little interest he ever had in questions like that, questions that aren't susceptible to being figured out. Broecker likes figuring things out. He likes, above all else, putting a new piece in the puzzle. That is the best fun, the deepest joy.

Science is a system, a way of thinking and acting, and a community that allows you to taste that joy, on your luckiest days. It is the belief that if we observe the world carefully, test our ideas skeptically, and communicate honestly, we can figure things out. That summer of 1952, Broecker was converted to science. In time he would come to think of it as something sacred.

THREE

Ice Ages and the Serb Theory

Sometimes when you look back on a life or a career, it makes a kind of sense—you get the impression it was planned that way from the start. Broecker has never been one to do a lot of planning. Even today he likes those days best that are the least scheduled, that contain plenty of room for something unexpected to happen. And yet reading today the PhD thesis he completed half a century ago, in 1957, you might think he had an overarching plan all along. All the threads of his later work are there, in separate chapters; they're just not woven together yet. Larry Kulp urged him to weave them together. In a final chapter, he suggested, young Broecker might explain why ice ages happen. Back then, climate science too was young, graduate students could run labs, and it seemed reasonable to ask them to solve century-old foundational mysteries. Broecker wasn't ready to write that last chapter in 1957. He has spent much of his career trying to write it. Until you understand ice ages, he figures—the biggest environmental change our planet has known in the past few million years—you haven't understood climate.

But in those early years Broecker put a big piece in the puzzle. When he radiocarbon-dated the tufas he had brought back from several expeditions with Phil Orr, he found that Lake Lahontan had reached its last highstand around 11,500 years ago. So had Lake Bonneville, the vast ancestor of what is now the Great Salt Lake. Around 11,000 years ago, according to Broecker's measurements, both lakes dropped sharply, receding into their present basins within a few centuries.

More striking still, that date, or something close to it, kept popping out of his radiocarbon counter when he put completely different samples in it. Oil-company geologists had sent him sediment cores from the Mississippi River, which also recorded a sharp transition at the end of the last ice age; once a torrent that funneled meltwater from the continental ice sheet to the Gulf of Mexico, it became the lazy meanderer we know today, depositing fine-grained mud on its bed instead of sand and gravel. That change, Broecker found, also happened around eleven thousand years ago. Meanwhile, a Lamont colleague, paleontologist David Ericson, had decided he could see the last ice age—and several previous ones—in sediment cores that Ewing's team had collected from the tropical Atlantic. As Ericson scanned down those cores, deeper into the past, he discovered several layers with a conspicuous dearth of shells of the warm-water planktonic species *Globorotalia menardii*. G. *menardii* is abundant in the tropical Atlantic today, and its bulbous but tiny shells, less than a millimeter wide, litter the top layer of seafloor mud. Broecker dated the reappearance of G. *menardii*, which according to Ericson marked the end of the last ice age and the return of warm water. The dates Broecker got from many cores clustered around eleven thousand years ago. Apparently the Ice Age had ended in the Atlantic at the same time as in the Great Basin and the Mississippi Valley, and just as abruptly.

Sometimes when you look back on how an important scientific truth got established, it doesn't make logical sense at all; too much luck and intuition were involved. Decades later, Broecker would learn from other scientists, working with improved dating techniques, that his eleven-thousand-year date wasn't quite right. Glenn Jones, then at the Woods Hole Oceanographic Institution, showed that G. *menardii* actually reappeared in the Atlantic less than seven thousand years ago; worms churning the seafloor had blurred that boundary in the sediment, mixing a few shells into the deeper, older layer that Broecker had dated. Lake Lahontan really reached its last highstand, it is now estimated, around fifteen thousand years ago; Broecker's tufas had apparently been infiltrated by rainwater and contaminated by younger carbon, and it has also been learned that a correction must be made to all radiocarbon ages to convert them to calendar years. Broecker has never gone back to check how his dating of Mississippi sediments has stood up. It doesn't matter much: he already knows it was pure luck that produced the coincidence of dates that led to the main conclusion of his thesis.

And yet later research has also proved that conclusion correct: the Ice Age really did end abruptly, if not at the precise date Broecker identified. After being gripped by cold for seventy thousand years or more, Earth came out of it in a few millennia—a geological instant. One instant, mile-thick ice sheets covered most of Canada and Scandinavia, reaching south as far as Chicago and New York, London and Berlin, and the next instant the ice was gone. One instant the Great Basin was a landscape of pine forests and vast lakes, and the next instant it was a desert. It was a huge and almost unimaginable change in climate, all the more difficult to understand for its rapidity. Half a century after Broecker first wrestled with the Ice Age, neither he nor anyone else fully understands it. In the meantime, what started as an academic puzzle has become a problem with much higher stakes: if we are to foresee the future of our climate, we need to understand better that crucial episode in its past.

One of the first people to imagine the Ice Age, and one of the most curiously influential, was a Swiss carpenter, mountaineer, and chamois hunter named Jean-Pierre Perraudin. If you go to the village of Lourtier in the Val de Bagnes, southeast of Lake Geneva off the road to the Grand Saint Bernard Pass, you can still visit the cozy chalet he built and spent his entire life in, after he returned in 1800 from fighting in the army of Napoléon. Perraudin's great-great-great-grandson was born in the house, and he and his wife now maintain it as a museum. On one of the pine-paneled walls hangs an early daguerreotype that may be a portrait of Perraudin (the provenance is a bit uncertain). It shows an old but still sturdy-looking rustic wearing two frayed jackets and an equally frayed and floppy hat; keen eyes flash out from a weathered, stubbly face. Perraudin himself carved what appears to be a crude self-portait, a bit

Paleolithic in style, into one of the stout ceiling beams in his living room. It shows a hunter pursuing a chamois.

In those days, in the early nineteenth century, it was mostly chamois hunters and their ilk who ventured into the high Alps, beyond the limits of summer pastures and to the edges of the glaciers; the age of tourism had not yet really begun. On his way to bagging a career total of 184 mountain goats, Perraudin had the time—and the eyes and mind—to notice some subtle and not-so-subtle features of the Alpine landscape. The subtle features were linear striations scratched into rocks on the sides and floor of the Val de Bagnes; the lines always seemed to be oriented along the axis of the valley. The not-so-subtle features were the giant angular boulders scattered around the valley, some teetering precariously atop other rocks, some perched high up the slopes. They were made of granite from the surrounding mountaintops. More eminent naturalists had noticed similar boulders in other Alpine valleys and called them erratic blocks. The most common idea was that they had been transported to their erratic positions by the Flood.

Perraudin believed in the Flood too, but he didn't believe it or any lesser deluge could have carried chalet-size boulders as far as Martigny, twenty-five miles from the head of the Val de Bagnes. Nor did he see how it could have carved those deep striations in other rocks. But standing next to one of the glaciers at the head of the valley, and seeing its enormous mass with its swept-up inclusions of rubble, he could picture it etching striations in rocks as it ground over them; and he could imagine it carrying other rocks, intact with their angles unrounded, to far-off places. Perraudin knew from experience that glaciers could in fact advance. These ones had visibly advanced in his lifetime. Like other glaciers in the Alps—like glaciers all over the world, climate scientists now realize—the ones above the Val de Bagnes were just then, in the early nineteenth century, on their way to their greatest extent in historical times. It was the last gasp of a cold period now called the Little Ice Age, which stretched from the fourteenth to the mid–nineteenth century.

In early 1818 the advance of one of the Val de Bagnes glaciers, the Giétro, caught the attention of the world outside the valley. A series of hard winters had brought the front of the glacier to the edge of a steep slope, and great blocks of ice had broken off and crashed more than two thousand feet to the valley floor. Shattering and then refreezing into a solid mass, they eventually closed off the narrow valley. In the spring of 1818 the Dranse River, normally a torrent rushing through Lourtier and the other villages in the valley, ran dry. To the horror of those who hiked up to the head of the valley to investigate, the river was imprisoned by the ice dam in a lake that was already more than two miles long and two hundred feet deep. It required no great insight to foresee a disaster once summer started to do its work on the ice dam. That spring the canton of Valais sent its chief engineer, Ignace Venetz, to look into the matter. Venetz immediately launched an urgent effort to tunnel through the ice dam and allow the lake to drain gently—but in vain. On June 15, a frightened Venetz spent the night awake on the glacier, listening to the cracking ice. The following afternoon, the dam suddenly burst. In half an hour, 18 million tons of water roared down the valley, preceded by a wall of trees, mud, and shattered houses. Venetz, who was running down the valley to warn its inhabitants, watched from a safe height as the flood overtook him. Some five hundred houses were swept away; debris floated on Lake Geneva. Some fifty people died.

As far as the history of climate science goes, however, the crucial event was not the disaster itself but the meeting it triggered between Venetz and Perraudin. The two men worked together in both the run-up to and the aftermath of the disaster—Perraudin often served as a guide for visitors to the valley—and they had plenty of reason to talk about advancing glaciers. One imagines them gathered around the cylindrical stone hearth in Perraudin's chalet, his rifle propped in one corner, candlelight flickering on the dark pine walls. "The glaciers in our mountains were once of far greater extent than today," the mountain man told his learned guest. "Our entire valley was occupied, to a great height above the Dranse, by a vast glacier that reached as far as Martigny . . ." Perraudin, who was a bit of a nut about his glaciers, had tried to sell this same story three years earlier to another scientific man, Jean de Charpentier, director of the salt mines at Bex; Charpentier, a friend of Venetz's, had listened politely but had dismissed the idea as "extravagant." Venetz, for some reason, did not dismiss it.

Indeed, as his work took him into other valleys around Switzerland, he convinced himself that Perraudin's idea was not extravagant enough—that the ice had extended much farther even than Perraudin imagined. It was Venetz who first recognized the significance of moraines: the piles of jumbled rubble that glaciers accumulate and deposit in long arcs along their margins. He could see moraines forming at the fronts of glaciers, and farther down the valleys, he could see entirely similar arcuate hillocks. The moraines showed that the glacier that filled the Val de Bagnes flowed into a far larger and longer glacier that filled the entire Rhône Valley. Stretching well over a hundred miles across Switzerland, from east to west, the Rhône Glacier then veered north across the foreland of the Alps and deposited erratic boulders on the flanks of the Jura Mountains. The climate in Switzerland, Venetz reasoned, must have been quite a bit colder during that epoch. The glaciers were like thermometers that revealed the temperature of the past.

Venetz persuaded Charpentier, who was initially horrified by his friend's strange enthusiasm; Charpentier persuaded Louis Agassiz, who had visited his friend Charpentier with the intention of talking him out of his folly; and Agassiz, the superior communicator of the three, persuaded the world that an ice age had really once happened. All that persuasion took a while. Agassiz started the boulder rolling in 1837, at Neuchâtel, when he delivered an address to the Swiss Society of Natural Sciences, of which he was president. He had been supposed to talk about his specialty, fossil fish, but changed his topic at the last minute. (The phrase *sex appeal* was not yet current, but some scientific topics already had more of it than others.) Before his startled audience, Agassiz did not hesitate to surge boldly beyond the evidence he had gleaned from Venetz and Charpentier. His ice cap, instead of advancing modestly from the Alps into the surrounding countryside, reached from the north pole all the way to the Mediterranean, extinguishing all life in Europe, much like the Flood. It had a similar extent and similar effects in North America and Asia.

Yet in the decades that followed, as scientists on both sides of the Atlantic were slowly converted to Agassiz's outrageous theory—Agassiz himself carried the news to America, becoming a professor at Harvard in 1847—and as they began to comb the landscape for the traces of vanished ice sheets, Agassiz's initial conjecture turned out to be not nearly as wild a departure from the truth as the truth was from people's ice-free experience. The European ice sheet never reached the Mediterranean, but it did cover Scandinavia, Britain, and northern France and Germany; south of it, a broad belt of forests was replaced with a desert of windblown dust. In North America, moraines turned up as far south as central Illinois—once people knew to look for them. Imagine an American scientist, however ingenious, standing in Illinois in the nineteenth century and, without the Swiss as intellectual Sherpas, spontaneously hatching the idea that the gentle hill in front of him was the work of a vanished, mile-thick ice sheet, whose roots lay twelve hundred miles north, in Hudson Bay. It couldn't and didn't happen that way. Probably the only place the Ice Age could have been recognized at that time was in the Alps, where people with scientific training—or at least a scientific mind, such as Perraudin—came into contact with glaciers.

Today the Alps remain one of the best places to try to get a feel for the Ice Age—and for why it has exerted such a fascination on scientists for two centuries now. Getting to the Giétro Glacier is strenuous if you don't have the conditioning of a chamois hunter, but you can drive right up to the edge of the Rhône Glacier, at an altitude of around seventy-six hundred feet in the Furka Pass. For a modest fee you can even walk *into* the Rhône Glacier; the owners of the Hotel Belvedere, which stands right next to it, have cut a cave into the ice every summer for more than a century. If you get there before the first tourist bus of the day, you can stand alone in the dripping cave and, with no one there to think it strange, press your eyes right up to the ice. The ice filters out red light, so it is blue—"an indescribable, almost achingly beautiful blue," in the words of Penn State glaciologist Richard Alley, who has spent many months on and in the Greenland ice cap. Looking into that three-dimensional blue yonder is like looking into the sky; trapped in the ice are bubbles of ancient air that are white, like clouds.

Outside the mouth of the cave, standing next to the front of the glacier, watching and listening to its meltwaters cascade sixteen hundred feet down a steep slope into the braided channel of the Rhône, you can look at the real sky and try to picture it as it once was: filled with ice. A recent doctoral dissertation by Meredith Kelly, now a research scientist at Lamont, helps anchor that reverie in hard fact. Kelly, an athletic young woman who looks fit enough to track a chamois, hiked hundreds of miles through the side valleys of the Rhone, in the footsteps of Venetz, Charpentier, and Agassiz, looking for traces of the vanished ice—especially trimlines. Whereas a moraine shows how far a glacier extended, a trimline shows how high up a mountainside it reached; the rock below the trimline is visibly polished and scored by the glacier, while the more jagged rock above the trimline was never ice-covered and was weathered only by wind and frost. Where Kelly couldn't find trimlines, she looked for roches moutonnées—bedrock hills that were rounded and polished by ice—or rat tails. A rat tail is a thin line of bedrock, a few inches to a few feet long, which escaped being eroded by the ice because it stood in the protected lee of something hard, such as granite; it points in the direction the ice was flowing, the way the still water behind a rock in a river tapers downstream.

Working with her adviser, Christian Schlüchter of the University of Bern, Kelly converted hundreds of such observations into a map of the undulating, flowing ice sea that covered this part of Switzerland twenty thousand years ago. The Rhône Glacier then was a dome of ice more than ninety-six hundred feet high; standing next to the waterfall at the front of the present glacier, you would have been under two thousand feet of ice. The ice flowed southwest down the Rhône Valley, but it also poured north over the Grimsel Pass into the Aare Valley, and northeast over the Furka Pass. Thirty miles down the Rhône Valley, at Brig, the ice surface was still a thousand feet higher than today's waterfall, which means the ice was around a mile thick.

At Brig, where Ignace Venetz attended a Catholic boarding school, the Rhône Glacier was broadsided from the north by the Aletsch Glacier, coming down a steep slope from the direction of the Jungfrau; Kelly and Schlüchter think the collision may have shoved ice south out of the valley, over the Simplon Pass and into Italy. A few miles farther down, a huge ice stream flowing out of the Vispertal, where Venetz was born and raised, overwhelmed the Rhône Glacier from the south, crushing it against the north wall of its valley. That stream drained the region around the Matterhorn, and it contributed most of the ice that continued west down the Rhône Valley. Later, at Martigny, it was joined by ice from Perraudin's Val de Bagnes, and probably from Mont Blanc. All this ice flowed together out onto the foreland north of the Alps, where it plowed into the Jura Mountains; flowed west beyond Geneva and east beyond Bern; and deposited, at various places, boulders the size of houses that hailed from various distant Alps. And all this detail rooted in careful modern fieldwork conforms more or less to the vision Ignace Venetz had 180 years ago, after Perraudin had put the bee in his ear. This crazy vision had made his friends ashamed for him, before it seized their imaginations too.

There is something so awesome about glaciers, and even more so about the glacial epoch—when the Alpine peaks we know today, judging from Kelly's map, were barren, windswept, sometimes frost-shattered islands poking out of a solid white sea—that people once exposed to them have a hard time putting them aside. Jean-Pierre Perraudin visited the Giétro Glacier upward of a hundred times, and clearly he wasn't just tracking big game. In 1819, a year after the Giétro triggered the flood that destroyed the church and sixty other buildings in Lourtier, washing away a lot of his own land and leaving him with large debts, Perraudin took a visitor to the source of his troubles. He found the glacier more beautiful than ever and congratulated his companion on having "the honor to contemplate it in all its glory." Ignace Venetz spent most of his later career on the practical work the cantonal government paid him to do—taming the Rhône River, draining swamps, and so on—but at the time of his death in April 1859, he was working on another manuscript about the Ice Age. It was a kind of memoir, recapping the genesis of the Ice Age hypothesis and all the evidence Venetz and others had since gathered for it. But it also broke new ground: Venetz suggested that the glaciers had advanced not just once but in four distinct epochs of the past. He laid all this out in thirty-four numbered sections, and then he continued:

> § 35.
> But what is the cause of these different extensions?

That was the last sentence he wrote; he had caught pneumonia in one of those swamps. Given the state of knowledge at the time, it was an appropriate place to stop. And a century later, when Broecker wrote his PhD thesis, it still was—but that was about to change.

In that same year, 1957, an old man whose great achievement lay decades behind him sat down in Belgrade to write, like Venetz before him, a short account of his life's work. Milutin Milanković was confident of his achievement. He had produced a mathematical theory that could, as he saw it, explain all major changes in climate, past and future, on this planet and all the others. In particular, he believed, the theory explained the recent ice ages on Earth. Milanković had an explanation too for why so many of his peers rejected his flawless theory: they were simply no good at math. They were empirical climate scientists, good at observing the world, perhaps, but ill-equipped to follow a "rationalist" such as him on his voyages beyond it, ill-equipped to understand the mathematical laws that govern its behavior. "Rationalists think best with their eyes closed, empiricists must keep them wide open," Milanković wrote at the beginning of his essay. He was trying to be charitable to his adversaries. Throughout his career he had repeatedly avoided answering them, not regarding it as his responsibility to fill what he saw as the yawning gaps in their mathematical education. But now, at the close of his life, he had to realize that the question of whether his work would last would depend on convincing the empiricists. So he resolved to try to present it, one last time, "in easily understandable form."

If you were to read only this one essay by Milutin Milanković, you would get an incomplete impression of the man. He was vain, brittle, and self-satisfied; he was a closeted scholar who thought he could fathom the universe without leaving his dusty, book-strewn office; he was obsessed with numbers. A certain well-known geologist named Penck had, Milanković wrote proudly, mentioned him fifty-eight times in a thirty-page paper. The Accademia dei Lincei in Rome had mentioned him fifty-six times during a scholarly discussion—Milanković couldn't read the Italian transcript, but he could pick out the word *Milanković*. To picture him bent over those papers, a balding man in a formal suit, his finger running down the pages to count the appearances of his own name, is to picture a nerd in a stuffed shirt. But Milanković was much more complicated than that. He was a precise man—"I do not like to think of things for which I cannot get an exact answer," he once wrote—but he was also a romantic; he was a methodical man who methodically constructed a theory of great power and beauty.

That other Milanković comes out in a book he wrote two decades earlier. Called, in the German original, *Durch Ferne Welten und Zeiten*—"Through Distant Worlds and Times"—it is a strong candidate for one of the strangest books ever written by a scientist. It consists of letters to a young woman in Vienna who had asked him to instruct her in the ways of the universe—she was rich, charming, and dainty, with chestnut hair and a sister in Norway, but Milanković's wife insisted she was merely a literary device. In any case, rather than subject her to dry scientific lectures, Milanković invites her along on imaginary voyages into space and the past, to places illuminated by science, so that she may feel the mystery of the cosmos, as it were, in living color. "Gray, dear friend, is all theory, and green is the golden tree of life!" he explains. And purple is the prose on occasion, but touching nonetheless, as the rationalist repeatedly digresses from the scientific business at hand to share autobiographical confidences. We see him break down in tears on his first visit to the Parthenon in Athens, whose geometry he already knows by heart—there is more to the world than is dreamt of in his mathematics, he finds. We see him feel the quickening pulse of his beloved, when he presses a furtive kiss into the buttonhole of her long, black, tight-fitting glove. "But this doesn't belong in my report," says Milanković, "it came to mind quite unexpectedly." As such things will.

More to the point, we hear the story of how in 1909 Milanković suddenly chucked a successful and lucrative career as a construction engineer in Vienna, as well as a failed love affair, for a professorship at the backwater University of Belgrade. "I wanted to be a real scholar and threw myself, like a fiery lover, into the arms of science," he wrote later. As a professor of applied mathematics in Belgrade, his job was to teach physics and celestial mechanics. When he went looking for a problem of cosmic significance to solve, the better to leave his mark on the world—he and a poet friend had adopted this goal one evening after splitting four bottles of wine—Milanković naturally hit upon a problem that lay at the intersection of his two academic disciplines.

More precisely, it lay at the intersection of two physical laws discovered more than two centuries earlier by Isaac Newton. The first was the law of gravitation, which says that the gravitational attraction between two celestial bodies increases with their mass and decreases with the square of the distance between them. That law was the foundation of celestial mechanics, which by then was a mature discipline; the sun's pull on the planets and even their pull on one another had been calculated with exquisite precision. The planets' mutual attraction tends to stretch and squeeze their elliptical orbits and to change the tilt of their spin axes. In the middle of the nineteenth century the French astronomer Urbain Leverrier had matched calculations of these effects with observations of Earth's actual orbit and had concluded that the calculated effects weren't enough—an unseen planet must be tugging on Earth along with the rest. Neptune was soon discovered at the exact time and place in the sky where Leverrier predicted it.

What no one had done, Milanković realized, was to pair this triumphant theory with Newton's second discovery, the radiation law, to create a theory of climate. Newton had showed how to calculate the amount of sunlight intercepted by a planet just as precisely as the gravitational force. Like gravitation, the radiation intensity decreases with the square of the distance from the sun, as sunlight fanning out in all directions gets spread over an increasingly large sphere. But unlike gravitation, the solar radiation delivered to a particular point on a planet's surface also depends on the angle at which it strikes—the sun is hotter at noon than at dawn or dusk.

So while the amount of sunlight reaching Earth as a whole doesn't vary much, the distribution does. It changes daily as the planet spins on its axis. It changes seasonally, because the spin axis is not perpendicular to the planet's elliptical orbit. Thus, during a year, first one hemisphere and then the other is tilted toward the sun and enjoys long summer days.

And the sunlight distribution must change too, Milanković knew, on the longer orbital cycles that Leverrier and others had calculated so precisely. There are three important ones. The spin axis, now tilted at about 23.5 degrees, oscillates between 22 and 24 degrees over a period of about forty-one thousand years. The shape of Earth's orbit, nearly circular right now, waxes more or less elliptical over a period of about one hundred thousand years. Both of those cycles are caused by the gravitational pull of the other planets, chiefly Jupiter and Saturn. The last one is different: it is caused by the pull of the sun and moon on Earth's bulging equator. That off-center tug causes the planet to wobble like a top, such that the spin axis, which now points at Polaris, the North Star, traces a circle on the sky every twenty-six thousand years. Other orbital variations interact with this precession to reduce the period of that sunlight cycle to around twenty-two thousand years.

All three orbital cycles modify the intensity of the seasonal cycle on Earth, and thus its climate. When the tilt of the spin axis increases, the summers get hotter and the winters get colder in both hemispheres, especially at higher latitudes; right now the tilt is relatively high and so the seasonal contrast is relatively strong. Meanwhile, as Earth's orbit gets more elliptical, the sun's position at one of the two foci of the ellipse gets more eccentric, which means that at one point on its orbit the planet is somewhat farther from the sun than at the opposite one. If that first point coincides with winter, the winter gets colder, but if it's summer, the summer gets milder—in other words, the distance effect harshens the tilt-induced seasons in one hemisphere and softens them in the other. It is the precession cycle, meanwhile, that determines which season coincides with which point on the orbit, by determining which way the spin axis is leaning. Right now, the northern hemisphere leans toward the sun and has its summer when Earth is farthest

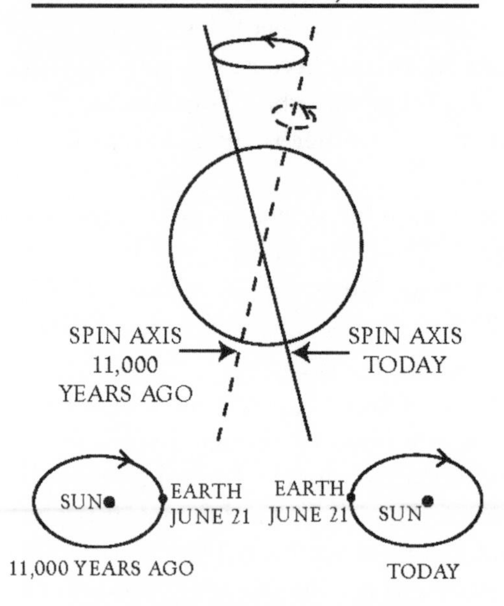

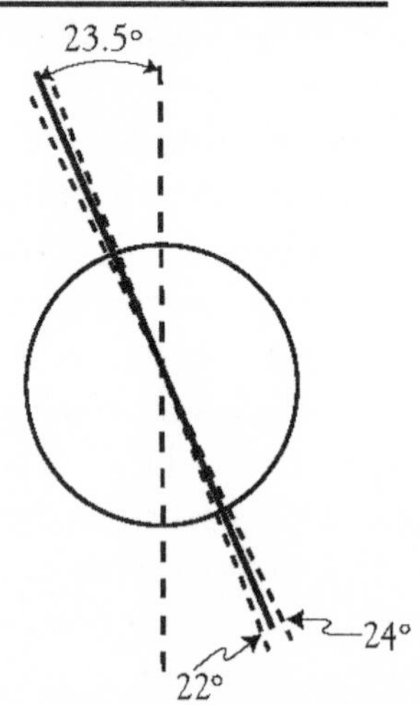

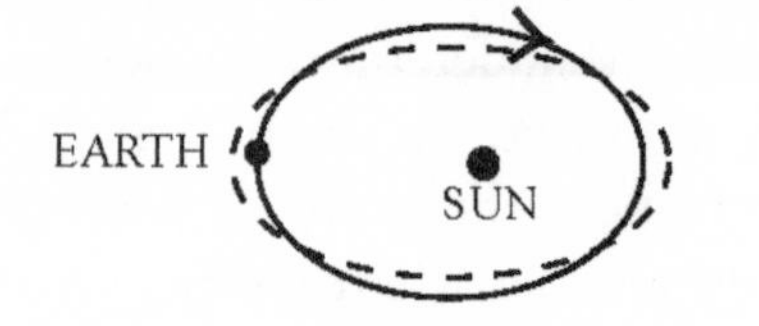

The Milanković cycles in Earth's orbit change the distribution of sunlight and thus the intensity of the seasons.

from the sun, in July, and the opposite is true in winter. As a result, the northern seasons are milder than they would otherwise be, while the southern seasons are harsher. But they're only slightly milder and harsher, because the orbit is nearly circular, and so the annual variation in Earth-sun distance is small.

All these cyclic phenomena were first identified in the nineteenth century, and various people, notably a brilliant Scotsman named James Croll, even put forward theories connecting them to the ice ages. What no one had done, Milanković realized, was to derive the formulas to calculate precisely what amount of sunlight the orbital cycles would conspire to let fall at a particular time and place on Earth, and what temperature that sunlight would produce. Such a theory, if he could construct it, would apply as far into Earth's past and future as he would care to extend the calculations, and to any other planet as well. (Indeed one of Milanković's first successes would be to calculate that Mars, whose supposed "canals" were just then capturing the public imagination, had an average global temperature of around one degree Fahrenheit—making it an unlikely residence for an advanced civilization.) The problem was exactly solvable in principle, that's what made it so attractive to Milanković, but it was not cookbook physics. You had to be a talented mathematician to derive the necessary formulas from Newton's fundamental laws, and you had to be bit of a maniac to then want to undertake years of calculations with those formulas. Milanković was both.

Neither war nor prison slowed him down; on the contrary. He started his project, which would take him three decades to complete, in 1911, and shortly thereafter he was drafted into the Serbian army as a staff officer in the war against Turkey. One of his first key breakthroughs on the climate problem came to him while he was watching his comrades struggle up a mountain to do battle. After the Turks were quickly defeated, and the Serbs attacked Bulgaria, Milanković was discharged and returned to Belgrade. He found the university deserted; most of the students and faculty were still under arms. For a whole year, as Milanković holed up in a small, book-filled room to begin his calculations, no one knocked on his door. He loved that.

Then in 1914 a young Serb in Sarajevo assassinated Archduke Franz Ferdinand of Austria, and the World War broke out. Austrian troops promptly attacked Serbia. They surprised Milanković on his honeymoon at the family estate in Dalj, eighty-five miles northwest of Belgrade. He was showing his young bride the place that meant so much to him, the place where he had first discovered his love and aptitude for mathematics; where the German governess, recognizing his brilliance, had tolerated his habit of sleeping late and taking no instruction until almost noon; where on a rowboat rocking gently on the Danube, which bordered his father's land, he had dreamed his first dreams of greatness. The Austrians ripped him from that place and threw him into a jail cell at Esseg. For a little while after the iron door slammed behind him, Milanković was shocked witless; then he remembered he had his briefcase with him. He resumed his calculations.

Milanković spent the World War happily working on his theory—he was sprung from jail after a few months when an old professor in Vienna convinced the Austrian army to let him take up residence in Budapest, at his own expense. By 1917 he had finished the rudiments of his theory. It showed how to calculate the amount of sunlight falling on a particular latitude at a particular season at a particular geologic epoch, and it also showed, to his satisfaction anyway, that changes in the sunlight distribution were large enough to have had large effects on climate in the past. It was not immediately obvious, however, how to apply this theory to explaining the ice ages. They were caused, one might think, by a diminution of sunlight, but at no time was sunlight diminished for all latitudes and all seasons—that's not how the orbital cycles work.

In the early 1920s an eminent and elderly German climatologist named Wladimir Köppen—who with his son-in-law Alfred Wegener, inventor of the theory of continental drift, was just then writing a book on climates of the past—read Milanković's first book and immediately got in touch. He suggested that Milanković calculate how sunlight had fluctuated over the past 650,000 years between the latitudes of fifty-five degrees and sixty-five degrees north—judging from moraines and other evidence, the edge of the continental ice sheets had most often been in that latitude band. And more specifically, Köppen urged Milanković to calculate how sunlight had fluctuated in summer. Summer, Köppen believed, was the crucial season for the growth of the ice sheets; in winter in the north it was always cold enough for snow to fall, but it would accumulate from one year to the next only if the following summer was too cool to melt it.

Köppen's theory was really only a plausible hunch. Some people disputed it then and some dispute it still; some argue that ice sheets form not when northern summers get cool, but when tropical winters get warm enough to evaporate lots of water off the ocean and send it north, where it falls as snow. But Milanković was not an empiricist, and he didn't worry about such niceties. He took the hunch and ran with it, just as if it were an empirical fact. For one hundred days he calculated from morning to night. Then he sent his results—his summer sunlight curves—to Köppen.

"The grizzled scholar compared these hieroglyphic markings with the documents of Earth history," Milanković later told his fetching literary device, "and found the course of the Ice Age faithfully represented in them." The "documents" in question were above all a diagram prepared by two German geographers, the aforementioned Albrecht Penck and his colleague Eduard Brückner. Penck and Brückner had decided that gravel terraces on the flanks of Alpine valleys had been deposited by advancing glaciers in four separate glaciations. By crudely estimating how fast sediments had accumulated since the last ice age, they estimated even more crudely the length of the whole Ice Age, also known as the Pleistocene Epoch, which consisted of four lowercase ice ages or glaciations. This was the 650,000-year figure that Köppen had passed on to Milanković.

Köppen put Penck's glacial-advance curve next to Milanković's summer-sunlight curve, eyeballed them both, and decided that the four glacial advances could be lined up, more or less, with minima in the amount of summer sun at sixty-five degrees north. That was enough for him—and when another study of river terraces near Weimar yielded similar results, it was more than enough for Milanković. The human brain, especially a rationalist kind of brain, is good at seeing what it wants to see, at deciding in the library or the closed office how the world should work—and then finding, in a confused jumble of conflicting natural evidence, the proof that the world really does work the way the brain knew all along that it must. By 1924 Milanković regarded his theory as proven. There was no need to find more geologic evidence that ice ages had happened when he said they did; henceforth, he thought, his orbital ice-age clock could tell geologists, who lacked a reliable timepiece, when their evidence had happened.

With hindsight this looks quite arrogant. With hindsight the evidence from the river terraces looks, even to the untrained eye, extremely slender. But it was worse than that: it was wrong. Those gravel terraces, it later turned out, weren't hard evidence of glacial advances after all; some had clearly been deposited in warm periods of no ice. One even contained a piece of a bicycle—which meant either that Cro-Magnon men rode bicycles across the glacier, or that the gravel had not been deposited by a glacier. And by the time Milanković wrote his valedictory essay, in the late 1950s, the first studies of ice-age formations with a new dating method—radiocarbon dating—were coming in. Those dates didn't seem to fit the orbital theory.

By the time Milanković died in 1958, his theory had fallen decidedly out of favor, especially in the United States. People were looking within the Earth system itself for the causes of the ice-age cycle. At Lamont, for instance, Maurice Ewing and William Donn were pushing a theory that placed the cyclic mechanism entirely in the Arctic Ocean, rather than in outer space. During warm periods, they argued, when the Arctic was free of ice, large amounts of water would evaporate from the sea surface and fall as snow on the neighboring continents, causing ice sheets to start growing. Those vast white expanses would then reflect so much sunlight back into space that the whole Earth would cool—and eventually the Arctic would freeze over again. This reflection feedback undoubtedly has an important impact on climate; Milanković himself had relied on it to explain how small changes in sunlight could cause such big changes in temperature on Earth. Unfortunately for the Ewing-

Donn theory, however, which is now generally disregarded, the evidence from Arctic sediments indicates the Arctic was not ice-free even during the interglacial periods.

When Milanković was a bright young boy in Dalj, his father had hoped he would study agronomy and lead the estate, which had been in the family for centuries, to new heights. That had never been ambition enough for Milanković. Instead, beginning with his happy confinement in Budapest during World War I, he had sold off the estate piece by piece to support himself and his science habit. In his last letter to the chestnut-haired device, he describes his last visit to Dalj, when he decides he must sell the house itself, provided his ancestors will grant their approval.

> I went to their graves. Bending over their place of rest, I whispered to them that in the search for knowledge and truth I had used up all the earthly goods that they had acquired and left to me. I asked them to forgive me and to consider, as a mitigating circumstance, that I had kept their name pure and through my work had built for it a more lasting memorial than the pillars and crosses on their graves.

His ego remained intact to the end—and yet Milanković was telling his ancestors the truth. The romantic rationalist would have the last laugh from beyond his own grave. In the years after his death, his orbital theory would enjoy a fate similar to that of the continental drift theory proposed by his friend Alfred Wegener: it would reemerge from the dustbin to gain recognition as one of the great insights into Earth science of the twentieth century. The key to proving Milanković right, or partially right—his theory is far from being a complete explanation of the ice ages, let alone of climate change in general—was to find an independent method of accurately dating geologic features of the ice ages, and to show they were deposited at the times predicted by the orbital theory. As it happened, dating was something Broecker knew a lot about.

FOUR

Proving Milanković, Doubting Milanković

Some of the first hard evidence linking ice ages to orbital cycles came from the island of Barbados, where the temperature today rarely descends below seventy degrees, and where even in the Ice Age it would have been balmy. The evidence was a couple of chunks of fossil coral. They arrived in the mail one day in 1967, at Broecker's Lamont office, and he unwrapped them eagerly. A few days earlier a geologist at Brown named Robley Matthews had phoned to ask whether Broecker could date the fossils. Matthews had money from Gulf Oil, he said, to investigate how the porosity of ancient coral formations changes over time; it's their porosity that makes them good oil reservoirs.

That wasn't Broecker's interest at all. Preparing himself a couple of years earlier to teach a course at Columbia in ice-age geology, he had steeped himself for the first time in the Milanković theory and had gotten inspired to test it in the field. Radiocarbon dating, Broecker's specialty, had been used to dispute the theory, but he knew it wasn't really adequate to the job. You couldn't find out whether Earth was responding to orbital cycles of twenty thousand or forty thousand or a hundred thousand years with a dating technique that took you back only thirty-five thousand years. Beyond that limit, the amount of radiocarbon in a sample had decayed to a level too small to measure—although people sometimes claimed to have done it anyway.

By then, Broecker's lab had moved beyond radiocarbon dating. A graduate student of his, David Thurber, had shown you could date coral fossils as old as two hundred thousand years by measuring their content of uranium and its radioactive daughter, thorium. A growing coral absorbs uranium from seawater, but not thorium, which settles out of the water too fast; over time the uranium decays to thorium at a known rate—much slower than radiocarbon—such that the ratio of the two, trapped inside the crystal matrix of the coral shell, falls like sand in an hourglass. Using that clock, Broecker and Thurber had already dated a few coral reefs in the Bahamas and elsewhere. They had shown that the ice age before the last one had ended, and sea level had reached a highstand, around when Milanković said it should have—about 120,000 years ago. One date, however, does not demonstrate a cycle. Broecker was thus hungry for more dates at the very moment when Matthews, pursuing a different agenda, called him about Barbados. Sure, he told Matthews, send me a couple of coral fossils.

Barbados is a special place, geologically speaking. It sits atop a tectonic pile-up: the zone where the plate carrying the floor of the Atlantic Ocean collides with and dives underneath the Caribbean plate, plunging into Earth's mantle. Some Atlantic material dives deep, gets heated and melted, and rises back to the surface to erupt at volcanoes; the Windward Islands just to the west of Barbados are the result. But some of the mud gets scraped off the Atlantic plate, the way a plow scrapes snow off a road. Barbados is the peak of such a mud pile, which has been growing gradually for many millions of years; it reached the sea surface and became an island sometime during the Pleistocene. At that point corals began to grow on it. Most of Barbados is capped with a thick layer of fossil corals. They are preserved in at least eighteen distinct reef terraces that ring the island and descend like steps from the highlands to the sea.

Those steps, Matthews and Broecker realized, are like a strip-chart recording of past sea level—and as such they're a far more accurate record of glacial periods than the gravel terraces studied in river valleys by Penck and Brückner. Whenever the continental ice sheets grew during the Pleistocene, removing water from the ocean and storing it in frozen form on land, sea level fell; when the ice sheets melted, the sea rose again. Each time the sea level stayed the same for a while, a reef terrace formed in the shallow water off Barbados, topped by staghorn and elkhorn corals that invariably grow right near the sea surface. In a more stable place, the only fossil terraces visible on land would be ones built when sea level was higher than it is today—in the last interglacial, for instance. But Barbados is rising out of the sea at the rate of about a foot a millennium. As a result, coral terraces that formed during relatively warm intervals of the last ice age, at sea levels that were relatively high but still lower than today's, have subsequently been lifted out of the sea. Matthews and his students could just walk up to them and hammer out staghorns and elkhorns that looked just like the ones growing today in the shallow water offshore.

When Thurber ran the uranium-thorium dates on the two fossils Matthews had sent, the first date he got, from a reef terrace that stood 130 feet or so above sea level on Barbados, was 124,000 years—more or less the same date he and Broecker had gotten elsewhere for the last interglacial. Thurber's second date, from a terrace closer to the sea, was 82,000 years. Broecker is an excitable guy when it comes to science, and those results excited him. He had done the math: 124,000 minus 82,000 was very close to 41,000 years, which was the length of the cycle in the tilt of Earth's axis—the cycle Milanković said was most important. The Serb had in fact calculated that there should have been a peak 82,000 years ago in the amount of sunlight reaching the high northern latitudes. Ice thus seemed to have melted and sea level at Barbados seemed to have risen in perfect synchrony with the tilt cycle. Broecker called Matthews and urged him to drop his coral-reef porosity work and climb aboard the Milanković bandwagon. Above all, he should send Broecker more samples.

Whereupon Matthews wondered aloud about a reef terrace he hadn't mentioned before, one that lay at an elevation halfway *between* the two that Thurber had dated. How did it fit into Broecker's beautifully synchronous cycles of sun and sea?

Broecker did not stay deflated for long. He had, in effect, already solved the problem. In plunging into the Milanković theory he had become aware of all the assumptions buried in it. Milanković had assumed that the amount of sunlight falling on high northern latitudes during summer was the key determinant of whether ice sheets advanced or retreated; and that had in turn led him to assume that the key orbital cycle was the 41,000-year tilt cycle, because the tilt of Earth's axis has the biggest effect on how much sunlight reaches the high latitudes. But the ocean and the atmosphere both spread solar heat from the tropics to the poles. Might they not flatten out the peaks and valleys of the tilt cycle? And might that not allow the precession cycle, which determines how far Earth is from the sun during a given season and thus affects the whole planet uniformly, to assume a greater importance? The precession cycle has a period of 22,000 years. When Broecker recalculated Milanković's sunlight curve, giving greater weight to the precession cycle, he got a new sunlight peak around 106,000 years ago. When Thurber subsequently dated the third terrace on Barbados, the one bracketed by the first two, he got a date for it of 105,000 years—close enough for paleoclimatic work.

Counting the end of the last ice age, eleven thousand years ago, which Broecker had dated elsewhere, there were now four well-dated rises in sea level that corresponded to peaks in Milanković's sunlight curve, as modified by Broecker. "The often-discredited hypothesis of Milankovitch must be recognized as the number-one contender in the climatic sweepstakes," Broecker and his colleagues announced in *Science*. But by then Broecker himself was already convinced that the orbital cycles were far from the whole explanation of how Earth's climate had swung in and out of ice ages during the Pleistocene. For one thing, the changes in the distribution of sunlight, only a few percent, seemed too small to cause such huge changes in climate, even aided by the positive feedback of changes in Earth's reflectivity. For another, the smooth sinusoidal variations in sunlight could not explain why the last ice age had ended not smoothly at all but abruptly.

There were more feedbacks in the air and sea, Broecker decided back then, in the late 1960s, than were dreamt of in Milanković's philosophy. Even before the Barbados results, Broecker had put forward the hypothesis "that the ocean-atmosphere system has two stable states or modes, glacial and interglacial; that rapid transitions between these states are triggered by the larger insolation peaks." Maybe all the orbital cycles did, in other words, was kick the great slumbering climate beast. And maybe they weren't the only thing that could kick it. But it would be another two decades before Broecker or anyone else would explore that hypothesis fully and consider what it implied, not just for the Ice Age, but for the future.

Broecker's papers on the Milanković theory helped bring it to wide attention in the United States; Milanković's key book was even translated into English for the first time. Broecker, always keen to get people interested in and working on problems he himself was interested in and working on, gave numerous talks on the subject. At one of those talks, an informal Friday-afternoon affair at Lamont, a good friend of his named John Imbrie was in the audience. Imbrie was chairman of the geology department at Columbia and only a few years older than Broecker; he had arrived in the same year, 1952. In matters of department politics they had formed a sympathetic alliance against older and more well-stuffed shirts. Once, on graduation day, when the two of them had just returned from the ceremony to Imbrie's office, the door opened and an old German gentleman in a dark suit walked in and clicked his heels. He was a distinguished oceanographer, Georg Wüst, just arrived on sabbatical, and preceded, fairly or not, by a reputation of being something of a Nazi. He was there, he announced, to present his credentials to the chairman of the Columbia geology department, whom he must have felt safe in assuming would also be a distinguished older gentleman. Broecker and Imbrie had their feet on Imbrie's desk. They still had their mortarboards on, and their splayed academic robes revealed shorts and bare legs. They looked at Wüst, and then at each other. "That's him," Broecker said finally, pointing at his friend.

Years later, sitting and listening to Broecker talk about coral reefs and orbital cycles, Imbrie felt almost as if his friend were pointing at him again. His specialty at the time was not the ice ages of tens or hundreds of thousands of years ago; it was the Paleozoic Era, which ended more than 200 million years ago. Imbrie had spent more than a decade tramping around places like Kansas and Michigan, knocking fossil clams and snails out of Paleozoic limestones, and trying to map the shallow seas where those animals had lived. It was earnest scientific work—but it had started to bore him. Broecker's talk did not. "I thought, 'Gee, that's more interesting than what I'm doing,' " Imbrie remembers. "And I began reading about the Milanković theory." Soon thereafter he got himself a chunk of sediment core from David Ericson at Lamont, took it with him to Scotland on a sabbatical, and made the switch from clams and snails to foraminifera, and from the Paleozoic to the Pleistocene. Within a decade he and his colleagues had extracted from that and other sediment cores the most conclusive proof yet that the Milanković theory, as far as it goes, is true.

Sediment cores were the key to verifying the theory, because compared to coral reefs or other records on land, they provide a continuous record of climate change. In 1955, around the time Broecker was discovering the ice age in the Nevada desert, a paper published in the *Journal of Geology* had pioneered the modern use of sediment cores; Broecker had first learned about Milanković from that paper. Called "Pleistocene Temperatures," it was written by a young Italian geologist at the University of Chicago, Cesare Emiliani. Emiliani too had gotten samples from Ericson—slices of mud cut at intervals of four inches from several cores, including a thirty-foot-long one from the Caribbean.

But whereas Ericson had scanned that same mud under a microscope, identifying the intricate shells of particular species of planktonic foraminifera that he considered indicators of either warm or cold water, Emiliani did something different: he crushed the foram shells. Then he washed the crumbs with distilled water, pulverized them in a mortar, and baked them at nine hundred degrees Fahrenheit in a stream of helium gas. From that perfectly clean powder of calcium carbonate, Emiliani extracted the oxygen the forams themselves had built into their shells thousands of years ago, when they were alive. With a mass spectrometer, he counted how many of those oxygen atoms were the light isotope, oxygen-16—which makes up more than 99 percent of all the oxygen on Earth—and how many were the heavier and much rarer isotope, oxygen-18. Normally the heavy oxygen prefers to be in calcium carbonate over water, because that reduces the overall vibrational energy of the molecular system. But as the temperature of the seawater goes up, that slight preference goes down, and with it the ratio of oxygen-18 to oxygen-16 in the foram shells. That ratio was the thermometer Emiliani used to take the temperature of the Pleistocene Atlantic.

In the half century since Emiliani's pioneering study, the measurement of isotope ratios has become one of the main tools of all research into ancient climates. But the interpretation of the oxygen-ratio results has changed—and that was one of the breakthroughs that led to the proof of the Milanković theory. Emiliani thought he could see Milanković cycles in his Atlantic and Caribbean sediments, but he hadn't convinced many people. His dating of the sediments, and thus of the coincidences between his temperature variations and the orbital variations, was based on a single radiocarbon measurement in the young sediments at the top of each core; from there he extrapolated through hundreds of thousands of years of mud by assuming that it had always accumulated at the same rate. And the isotopic temperature measurement itself was unconvincing to many researchers. It contradicted the testimony of the forams themselves. After he got back from his Scottish sabbatical, Imbrie developed a much more sophisticated statistical version of Ericson's technique for estimating past seawater temperatures; instead of using individual foram species as indicators, he looked at how whole assemblages of species changed over time. That allowed him to distinguish changes in temperature from changes in other environmental factors, such as the salinity of the seawater, that might also affect forams. According to Imbrie's equations, the surface of the tropical Atlantic had been no more than two degrees Celsius colder during the Ice Age—whereas Emiliani's isotope measurements had put the chill at six degrees.

The problem, as Nicholas Shackleton at the University of Cambridge showed, was that oxygen isotopes record something other than just the temperature of the water that bathed the forams: they record the volume of ice all over the planet. Shackleton had the bright idea of comparing the isotope ratio of forams that had floated at the sea surface with that of forams that had lived on the seafloor at the same time (and were thus preserved in the same layer of mud). Even at the height of the Ice Age, he reasoned, the bottom of the ocean could not have been much colder than it is today, for it is already near freezing. And yet the bottom-dwelling forams yielded pretty much the same oxygen-isotope curve as the planktonic ones. During the coldest periods of the Ice Age their shells were relatively rich in the heavy oxygen isotope and poor in the light; during warm periods it was the opposite. Since the temperature of the water at the seafloor couldn't have changed that much, something else must have—the isotopic composition of the water. When water evaporates from the sea, H_2O molecules that contain the lighter O isotope evaporate more readily than the heavier ones, and the heavier ones are also more likely to condense again and rain right back into the sea. During a glacial period, when a huge volume of water snows onto the land and gets locked up in ice sheets, the whole ocean gets enriched in heavy oxygen and depleted of light oxygen.

Emiliani, though aware of this effect, had underestimated its importance—but the truth was that if his oxygen-isotope ratio was recording the amount of ice on the whole planet, rather than the water temperature at one particular spot, it became an even better indicator of Ice Age climate. Technical as that ratio is, abstruse as it is, when you think for a minute about what Emiliani and Shackleton discovered, it is astonishing and beautiful. A foraminifer shell is the size of a grain of sand; on the beach it *looks* like a grain of sand. And yet the oxygen atoms in such shells at the bottom of the sea record the advance and retreat of mile-thick ice sheets across the face of the continents, and thus the complete alteration of our world that has taken place at regular intervals over the past million years. The whole planet in a grain of sand—it seems too poetic to be a scientific truth. But it is true, and when Shackleton announced his results at a meeting at Lamont in 1972, Imbrie and others in the audience wanted to cheer. "We were sitting there and Shackleton announces this, and immediately everybody says, 'By God we've got it,' " Imbrie recalls. "We've got like a golden key."

The researchers in that room, geologists and paleontologists mostly, had been assembled by Imbrie and Jim Hays of Lamont; they were all working on a project called CLIMAP, the goal of which was to make a climate map of the world at the height of the last ice age. To that room full of stratigraphers, Shackleton's discovery meant they could use the global oxygen-isotope curve to line up sediment cores from different parts of the globe and identify the strata of mud in each one that had been laid down at the same time. Compared with Emiliani working twenty years earlier, they had a huge advantage: they could fix a firm date to the bottom of their cores as well as the top. In the early 1960s, geologists had confirmed that Earth's magnetic field had spontaneously reversed its polarity, with north becoming south, on repeated occasions in the geologic past. The most recent reversal had occurred 780,000 years ago, according to radiometric dates of lava flows on land, obtained by measuring the decay of potassium to argon. A Lamont researcher named Neil Opdyke had located that reversal in the same sediment core from the western Pacific that Shackleton had analyzed, twelve meters deep in the mud. With that firm date at the bottom and the firm radiocarbon date near the top, the CLIMAP researchers could date the many peaks and valleys in the oxygen-isotope ratio that lay between—knowing they reflected advances and retreats of the ice sheets—far more reliably than Emiliani had. The valley representing the last interglacial had an age of 124,000 years, the same one Broecker had found in corals.

The most recent peak, when the foram shells contained a maximum of oxygen-18, occurred at the Last Glacial Maximum, or LGM—which CLIMAP put at eighteen thousand years ago. Imbrie's "factor analysis," applied to forams or to another kind of plankton called radiolarians in the LGM layer of hundreds of sediment cores, gave an estimate of what the sea-surface temperature had been in each of those places. And that information could then be paired with observations on land. A young geologist named George Denton of the University of Maine had made it his life's work to map glacial moraines and trimlines, in the tradition begun by Ignace Venetz. When the CLIMAP group finally published its map in 1981, it included Denton's picture of the extent of the ice sheets.

But even before then, in 1976, Hays, Imbrie, and Shackleton had obtained the most important result from CLIMAP. It came from two sediment cores from the southern Indian Ocean, which were long enough to reach to that 780,000-year-old magnetic

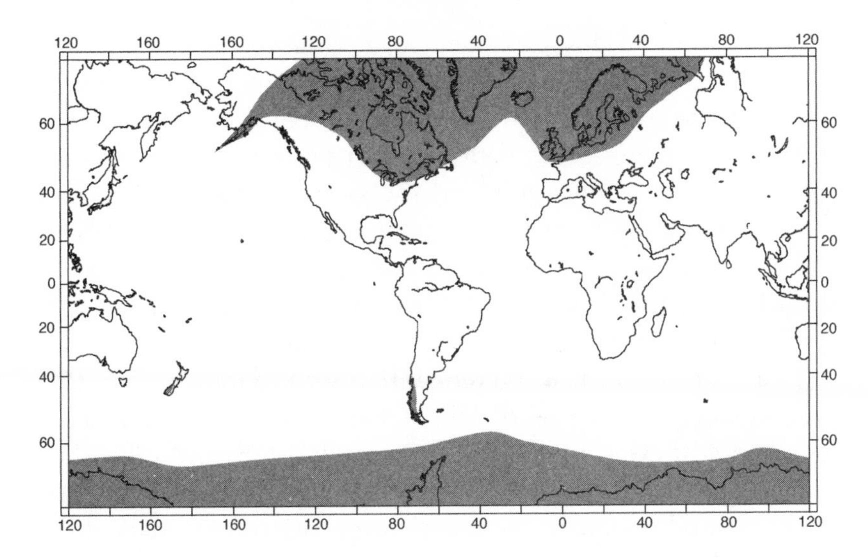

At the Last Glacial Maximum, ice sheets covered much of North America, northern Europe, and the continental shelf around Antarctica.

marker, and in which sediment had accumulated fast enough to provide hope of seeing the orbital cycles. Shackleton measured the oxygen isotopes in forams. Hays factor-analyzed the radiolarian populations to get sea-surface temperatures. And Imbrie, using a statistical technique called spectral analysis, extracted from those complex wiggling curves the dominant cycles that composed them—the way, he says, a complex piano chord is composed of a number of individual notes of specific frequency. Earlier Imbrie had shown that the 22,000-year precession cycle, the one Broecker had found to be so important, was itself composed of two subcycles of 19,000 and 23,000 years. Now, spreading out the printout from his mainframe computer on the table in the basement of his home outside Providence—he had left Columbia for Brown—he saw that the oxygen-isotope curve contained cycles of 42,000, 19,000, and 23,500 years. "Very close," he says. "I knew right away that this would not have occurred by chance alone. I thought, 'My golly, Milanković is right.' " The paper he and his two colleagues went on to publish in *Science*, which would become as influential as Emiliani's 1955 paper, was called "Variations in the Earth's Orbit: Pacemaker of the Ice Ages."

And yet the most prominent cycle Imbrie found in the sediments was not readily explained by a variation in Earth's orbit. That cycle was one hundred thousand years long. Broecker had called attention to it in 1970, when he reexamined some of Emiliani's own oxygen-isotope data. The last five ice ages, he pointed out, all seemed to have followed the same "sawtooth" pattern: a slow buildup of ice over ninety thousand years or so, followed by an abrupt "termination" that lasted less than ten thousand years. Hays, Imbrie, and Shackleton had shown that the Milanković theory could explain the timing—but not the magnitude—of the fits and starts of ice advance and retreat that seemed to punctuate that slow buildup. But the theory did not seem to explain the hundred-thousand-year-cycle itself—the eccentricity of Earth's orbit does vary with more or less that period, but the effect on sunlight is weak. Nor did it explain the abrupt terminations that marked the end of each glaciation.

And there was a third thing too. The two sediment cores that Hays, Imbrie, and Shackleton studied came from deep in the southern hemisphere. They suggested for the first time that ice-age changes in climate had happened simultaneously in both the northern and southern hemispheres. It was hard to see why that should be, considering only the Milanković theory. Why should changes in the amount of sunlight at a latitude of sixty-five degrees north have such a big impact on the temperature of the southern ocean or on the glaciers of New Zealand? Thirty years later, researchers are still trying to answer that question.

Outside the little town of Methven, on the South Island of New Zealand, the road heads northwest across the Canterbury Plains, toward the Southern Alps and away from the Pacific. It traverses cow pastures and sheep meadows separated by windbreaks of California pine. After a few miles of rising almost imperceptibly, it begins to rise a bit more noticeably, if you're the sort of person who notices that sort of thing. Then, after a little curve, it takes a slight dip. Immediately George Denton pulls the car over. Sheep and magpies in the neighboring meadow glance up from their work. "See?" Denton says. "You would have driven right over it." That dip in the road, he explains, was the face of a moraine constructed at the Last Glacial Maximum by an enormous glacier surging out of the distant mountains with titanic rock-pulverizing and world-shaping force. A shame you missed it.

We live our lives, many of us, on a landscape constructed by the Ice Age, and we don't see it. Broecker as a boy drove his bike and his Soap Box car over the bottom of a glacial lake and never knew it. His pal Denton grew up in and around Boston—you hear that right away when Denton says the word *gawgeous* (as in "This is a gawgeous moraine complex!"). Such Boston landmarks as Bunker Hill and the harbor islands are drumlins—streamlined gravel hills deposited by the advancing ice sheet—but as a youth Denton was not aware of that. He walked many times across the Wildwood Flats, near the family home in Winchester, without realizing that it, like the Canterbury Plains, or the North German Plain, or a lot of the American Midwest, was a glacial outwash plain, a flat stack of sediments deposited by meltwater rushing off a glacier, bursting through its previously constructed moraine, and winding to the sea through an ever-shifting series of braided channels. Denton as a boy was less interested in such things than in spending time at Fenway Park, where he claims to have done a lot of his homework. In his freshman year at Tufts, however, he walked on a whim into a geology course taught by a man named Bob Nichols. Nichols took Denton on a field trip to those same Wildwood Flats and made Denton himself decipher what they were. "Took me a while to figure it out," Denton says. "But I haven't been fooled since."

Denton is a tall, powerfully built man, a John Wayne kind of a man (except for the accent), with joint pains today that make him shamble like the older John Wayne, and with thinning hair that is more gray than the original reddish blond. In Alaska, where he spent most of the 1960s mapping moraines in the southern Yukon territory, and where he had a series of experiences with grizzly bears, Denton rode horses quite a bit. Helicopters were harder to come by then, and you forded the swollen rivers on a horse because the moraines were on the other side. In Antarctica, where Denton spent thirty field seasons, which means he spent eighteen hundred nights in a tent ("one too many," he says), helicopters were essential. The National Science Foundation would load a chopper into the same C-130 plane that carried the geologists and drop the whole lot off in the Ellsworth Mountains or somewhere; from that base camp Denton and his team would fan out into the ice-free mountains by helicopter, looking for moraines and trimlines and the like. One time the helicopter tipped backward off a sheer cliff and plummeted a thousand feet or so while a flustered young pilot went rigid with fear, and Denton, his equally large student Tom Lowell, and his frequent moraine-hunting companion Bjorn "Hound Dog" Andersen sat bug-eyed in the backseat—Denton still remembers the feel of Andersen's fingers, steely with terror, digging into his thigh. Anyway, they all came out of it alive; a more experienced pilot grabbed the controls in the nick of time. And these days Andersen, in his eighties now, short and bald and a creature of habit, still leaves the moraine he lives on in Norway every January and February, forsaking his pastime of cross-country skiing, to come map moraines with Denton in New Zealand. After all, when it's winter in Norway, it's summer in New Zealand, and summer in New Zealand is quite pleasant. And hunting moraines, and seeing the concealed history emerge from a landscape, is just plain fun.

Denton had already been doing it for decades in Antarctica and Alaska when he went with Broecker to the Andes in 1988. With money from Exxon, Broecker had invited two dozen geologists from both hemispheres on a field trip to examine the evidence for ice ages in Chile and Argentina—to see if the hundred-thousand-year cycle was truly global, and to think about what might have caused it. On the first night of the field trip, Broecker roomed with a Czech geologist named George Kukla. The two men had first met at a scientific conference in Paris in 1969 (more precisely, at a bar near the conference hall). On comparing notes, they had soon realized that each had independently discovered the same hundred-thousand-year cycle—Broecker in deep-sea sediments, Kukla in layers of windblown dust, or loess, in a brickyard in Brno. Each of those loess layers represented a time when the Scandinavian ice sheet had surged south into central Europe, transforming the broadleaf forests of Czechoslovakia into a windswept polar desert.

A couple of years after that meeting, Broecker had helped Kukla and his family defect from behind the Iron Curtain and come settle at Lamont; Broecker had picked them up at Kennedy airport. He and Kukla were thus old friends and close colleagues—but not, as it turned out, ideal roommates. On that first night at the hotel in Buenos Aires, Broecker retired early, exhausted from the long trip. Kukla, a compact bear of a man, rolled in a few hours later and almost immediately commenced a window-rattling snore. Broecker retreated to a couch in the hotel lobby, and the next night moved his bag to Denton's room. It was the beginning of another beautiful friendship, and of a scientific collaboration that continues today—though Broecker and Kukla also remain close friends.

Broecker convinced Denton to move his fieldwork out of Antarctica into the midlatitudes of the southern hemisphere. New Zealand in particular is a good spot to consider the question of whether the ice ages came and went on the same schedule in the southern hemisphere as in the northern. The Southern Alps, built by the collision between the Australian tectonic plate and the Pacific one, include a series of 10,000-foot peaks; the tallest, Mount Cook, is 12,500 feet high. Those peaks are less than fifty miles from the Tasman Sea, and the prevailing winds regularly dump huge amounts of snow on them—as much as fifty feet a year on some slopes. The result is permanent glaciers that flow downhill, etch rock, build moraines, and record climate changes just like glaciers in the Swiss Alps, which are close to the same latitude on the other side of the world. "This is a unique place—a small island standing right up in the middle of the southern hemisphere westerlies, way out in the South Pacific," says Denton. "It's remote from the place where all the hypotheses originated—the North Atlantic. If you want to see whether the changes are global, you come here."

And the best place of all to see them, the place where they are readily discernible even to the undiscerning eye, is the area around Mount Cook. Approaching it from the east and the Canterbury Plains, you pass Lake Tekapo, a long basin scoured by an ice-age glacier. It is now filled not with ice but with meltwater. Next comes Lake Pukaki, a ten-mile-long lake formed by glaciers flowing down the eastern flank of Mount Cook. Between the two lakes, the road runs through a field of what appear to be giant, grass-covered moguls, a hundred feet tall or more and dotted here and there with large boulders. Those are erratics, and this is a gorgeous moraine complex, Denton explains—one that records the restless advance and retreat of the glacier spilling out of the Pukaki basin east toward Tekapo. A glacier is like a conveyor belt, always moving downhill, always transporting rock and debris to its edges, where the melting ice drops its cargo; if that edge stays in one place for an appreciable time, the debris piles up to build a moraine. There are several ways of determining how old the moraine is. The latest and most accurate involves measuring isotopes produced by cosmic rays in the rock.

Denton and his team have spent the past decade identifying and dating moraines all over the Southern Alps and recording their results on detailed maps. The maps reveal not only where the ice was at a given time but also, more indirectly, what the temperature was. The Southern Alps, Denton says, are like an inverted dipstick: the way a dipstick in your car measures the height of oil in your reservoir, the glaciers of the Southern Alps measure the height of something called the equilibrium line. The equilibrium line is the altitude at which the two processes that shape a mountain glacier—the accumulation of snow high on the mountain, and the melting of ice lower down—are exactly in balance. It's also called the snow line, because at the end of the summer there is still fresh, unmelted snow on the ice above that line. The snow line moves down the mountain as climate cools and the glacier advances, and up the mountain as climate warms and the glacier retreats. From the location of a moraine you can determine the past shape of a glacier, from the shape you can determine the altitude of the snow line, and from the snow line you can calculate the prevailing temperature. Snow lines are accurate paleothermometers.

At the height of the last ice age, the snow line in New Zealand was about 750 meters lower than it is today. That means the temperature was about 4 to 4.5 degrees Celsius lower—the same drop that is recorded in Europe and North America. And Denton and his colleagues have shown that the dates in the two hemispheres are the same too. Going from Lake Tekapo to Lake Pukaki you pass first through moraines that are about sixty thousand years old—apparently the glaciers streaming off Mount Cook reached even farther then than they did at the Last Glacial Maximum. As you move closer to Pukaki, you then enter a broad belt of moraines that formed, according to Denton's dates, between 26,000 and 17,600 years ago. Those are the moraines of the Last Glacial Maximum. LGM moraines in the Alps or around the Great Lakes date from the same period. Around 17,600 years ago, the ice age in New Zealand suddenly terminated; the glaciers withdrew back up the mountains, leaving behind meltwater lakes such as Pukaki and Tekapo, dammed by moraines, to mark their former locations. Lake Zurich on the north side of the Swiss Alps and Lake Garda on the south side formed at precisely the same time in precisely the same way. And a sudden rush of "light" oxygen back into deep-sea sediments at that time confirms that ice was indeed melting all over the world.

"The ice built up over the hundred-thousand-year cycle," says Denton, stopping the car. The eerily beautiful, toothpaste-blue slab of Pukaki has come into view at last, its water clouded with fine glacial silt. "It stood at the LGM for several thousand years, oscillating and depositing these moraines. And then, by magic, the termination happened. It didn't take more than a thousand years for the glaciers to recede back into the mountains. That to me is the greatest climate change of the last ice age."

It's a change the Milanković theory can't explain; it's a change no theory can yet explain, though there have been many theories, including one or two that Broecker has put forward. No computer climate model can reproduce the termination. But one thing all theories and models agree on: carbon dioxide is somehow involved. Carbon dioxide, unlike the orbital cycles, affects both hemispheres equally, because it mixes rapidly through the atmosphere, and it's concentration is more or less the same all over the planet. At the height of the last ice age, judging from bubbles of ancient air trapped in the ice of Greenland and Antarctica, the carbon dioxide concentration in the atmosphere was only 70 percent of what it was right before the industrial revolution—195 parts per million instead of 280 parts per million. (Today it is at 383 ppm and climbing at the rate of almost 2 ppm a year.) In fact the latest ice core extracted from Antarctica, analyzed by a European team called EPICA, shows that the carbon dioxide concentration has moved in lockstep with the ice sheets for the past 800,000 years, through more than six full cycles of glacial and interglacial periods. It was always low when the planet was cold, and high when the planet was warm, just as you would expect of a greenhouse gas.

In 1972 these past fluctuations of carbon dioxide were not yet known. The Milanković theory was moving toward its moment of triumph. The proof that orbital cycles paced the ice ages, even if they didn't cause them, was still a few years away, but the idea that there had been a regular alternation of glacial and interglacial periods during the Pleistocene had gained general acceptance, at least among paleoclimatologists. Some of them began to turn their eyes toward the future. That year Robley Matthews and George Kukla, just off the plane from Czechoslovakia, convened a conference at Brown. The topic was "The Present Interglacial: How and When Will It End?"

Even paleoclimatologists are influenced by the weather. The winter of 1972, as Kukla, Matthews, and J. Murray Mitchell noted in their introduction to the conference proceedings, was cold and snowy—and it seemed to fit a long-term global trend, one that Mitchell, a National Oceanic and Atmospheric Administration (NOAA) climatologist, had first called attention to. Since the 1940s the whole Earth had apparently been cooling. The evidence was not just thermometer readings. Snow was covering once snow-free areas of Baffin Island. Pack ice was impeding navigation around Iceland. Armadillos were fleeing southward out of the American Midwest.

Most important of all, the trend made disturbing sense, given the rhythm of glacial and interglacial that some of the scientists assembled at Brown had done so much to reveal. "Nobody was able to present any documentation of a warm interval lasting more than ten thousand years," Kukla recalls—a fact that was "of immediate concern," as Cesare Emiliani emphasized in his talk at Brown, "because the present warm interval has already lasted close to ten thousand years." None of the papers presented at the conference warned that an ice age was around the corner. But in a few centuries or a few millennia—"relatively soon" in a geologic sense—anything was possible.

Later, Kukla and Matthews even felt compelled to take that sobering information to the highest authority, and to juice it up a bit. In December 1972, they wrote a letter to President Richard M. Nixon, warning him "that a global deterioration of climate, by order of magnitude larger than any hitherto experienced by civilized mankind, is a very real possibility and indeed may be due very soon . . . The present rate of the cooling seems fast enough to bring glacial temperatures in about a century, if continuing at the present pace." The consequences, they went on, could include lower food production and an increase in "floods, snowstorms, and killing frosts." And the kicker, for a president at the height of the Cold War, was this: "The Soviet Union, with large scientific teams monitoring the climate change in Arctic and Siberia, may already be considering these aspects in its international moves."

When skeptics today complain, as they so frequently do, that the same climate scientists who now warn about global warming were once crying wolf about global cooling, there is thus a kernel of truth to it—but only a small kernel. In fact the science is not the same and neither are the scientists. When Kukla and Matthews wrote their letter, climate science was still in its infancy, and there was little published literature to justify the degree of alarm they expressed. There was just a thirty-year cooling trend and the dawning realization that another ice age was inevitable someday. That realization has been borne out—but the inevitable has since been pushed far into the future. Contrary to what the researchers at the 1972 conference believed, there have actually been warm interglacial periods that lasted much longer than the one we're in now. In fact the closest analogue to the current interglacial is the one that ended around four hundred thousand years ago; Earth's orbit was nearly round then, as it is today. That warm period lasted thirty thousand years. So it seems we have ten thousand or twenty thousand years before we need to worry about an ice age.

Today, on the other hand, the scientific papers documenting the reality and the mechanism of global warming fill bookshelves. Even in 1972 at Brown, a number of scientists—including Mitchell, who had discovered the global cooling trend—were calling attention to the countervailing effect of carbon dioxide from fossil fuels. And in 1975, as articles on global cooling were still appearing in *Newsweek* and *The New York Times*, Broecker was predicting that the cooling trend, part of a natural cycle, would soon come to an end. The natural cooling had been masking a warming caused by our CO_2, he argued in *Science*, and when it ended, the onset of global warming would be all the more dramatic. "We may be in for a climatic surprise," Broecker wrote.

It was one of the first loud warnings about global warming, and the data it was based on—an analysis of oxygen isotopes in the first core extracted from the Greenland ice cap, at a place called Camp Century—later turned out to be questionable; the temperature cycles they seemed to show have not been seen elsewhere. And yet in 1975, the cooling trend did indeed end, and Earth's average temperature began a long upward climb that continues today. The year 2005 was the warmest on record. Broecker's evidence may have been misleading but his intuition was right—and this time he would have preferred not to have been so lucky.

FIVE

Carbon Dioxide and the Keeling Curve

No one paid much attention to Broecker's warning in 1975; not his fellow scientists, and certainly not the public. No one had paid much attention either to Svante Arrhenius, eighty years earlier, when he first pointed out the possibility that human beings were warming Earth—which Arrhenius, unlike most scientists these days, thought would be a good thing. The Swedish physicist and future Nobel Prize winner had devoted a year of his life and a lot of pen and paper to calculating, latitude by latitude, just how much the planet is warmed by carbon dioxide. "I should certainly not have undertaken these tedious calculations if an extraordinary interest had not been connected with them," he wrote at the end of that year, 1895. The interest for Arrhenius was not the future of climate, any more than it would be for Milanković two decades later when he plunged into his own tedious calculations. The interest was explaining the past swings between glacial and interglacial periods. Milanković was still in high school in 1895, but others had already put forward the first astronomical theories of the ice ages. Arrhenius was unimpressed by them. To him, a change in the transparency of the atmosphere—that is, in the greenhouse effect—did the job better.

Arrhenius did not discover the greenhouse effect. It was the French mathematician Jean-Baptiste Joseph Fourier who first realized, in 1827, that a planet's surface temperature is set by a balance of two energies: the light it receives from the sun and the invisible heat it gives off—infrared radiation, we now call it, but to Fourier it was *chaleur obscure*, or "dark heat." The planet warms up until it is radiating as much energy back into space as it is receiving. (If there were an imbalance, Earth would over geologic time have become either much too hot or much too cold to support life.) Because the atmosphere is selectively transparent—it lets through most of the sunlight but blocks much of the infrared and sends some of it back down toward the surface—the planet's surface has to warm up more and emit more heat than it would if it were naked rock. In the 1850s, the British scientist John Tyndall showed by laboratory experiment that it wasn't the atmosphere as a whole that was blocking the heat, as Fourier had thought. The nitrogen and oxygen that make up 99 percent of the atmosphere don't absorb infrared radiation at all. It's the carbon dioxide and water vapor, which together make up less than 1 percent of the atmosphere, that warm Earth.

What if their concentration were to change? That was the thought that spurred Arrhenius to link the greenhouse effect to the ice-age cycle. A Stockholm colleague of his named Arvid Högbom had just recently put forward the novel idea that the amount of carbon dioxide in the air might have varied widely in the past. So much less carbon was in the atmosphere than was locked up in limestone and other carbonate rocks on land and at sea, Högbom reasoned, that relatively small geologic fluctuations—a period of increased volcanic eruptions, which belch out CO_2 among other things, or a period of enhanced carbonate deposition—could cause large changes in atmospheric CO_2.

Encouraged by this argument, Arrhenius set about calculating whether such changes could have produced an ice age. He took into account a crucial positive feedback: as CO_2 in the atmosphere goes up or down, so does evaporation and thus the amount of water vapor in the atmosphere. That tends to amplify the temperature change caused by the CO_2. Allowing for the water-vapor feedback, Arrhenius found that cutting the CO_2 concentration by a little less than half from its 1895 value would have reduced the temperature in the midlatitudes by four to five degrees Celsius. That would in turn have sent ice sheets surging into those regions—simultaneously in both hemispheres, as the geologic evidence even then seemed to require, and as no orbital theory could adequately explain. Conversely, a doubling of CO_2 from its 1895 level would raise temperatures between five and six degrees Celsius. And a tripling might warm the Arctic by eight or nine degrees Celsius, thus explaining fossil evidence that Earth had once, before the onset of the ice ages, been far warmer, such that tropical plants and animals survived in the Arctic.

Arrhenius's theory enjoyed brief interest among his peers, but then faded from view, not to resurface for decades. Today his claim that natural changes in carbon dioxide can change climate seems prescient. Based as it is on uncontested 150-year-old physics, it is accepted by the most zealous skeptics of man-made global warming. Some of the most striking evidence comes not from Earth but from our neighbors in space, Mars and Venus. Anyone who has in recent years watched on television the spectacular footage sent back by NASA's Mars rovers has seen what happens to a planet that lacks a decent greenhouse: Mars is a wasteland, where the temperature only just noses above freezing in the tropics, and where the global average is minus sixty degrees Celsius. Most of the planet's carbon dioxide was long ago washed out of the atmosphere and stored in carbonate rock, as on Earth—but unlike Earth, Mars has no active volcanoes to return carbon to the atmosphere and maintain the greenhouse. Venus, on the other hand, does have active volcanoes—what it lacks, because its ocean boiled away billions of years ago, is water to wash the CO_2 out of its atmosphere. The Venusian atmosphere is almost pure CO_2, in fact, and is sixty-six times as dense as Earth's. It's a "runaway" greenhouse: the average surface temperature on Venus is around 450 degrees Celsius, nearly 850 degrees Fahrenheit—hot enough to melt lead, hot enough for rocks to glow, and hot enough, in combination with the intense atmospheric pressure, to disable the series of Russian spacecraft that landed on the planet in the 1970s and 1980s. They typically lasted less than an hour.

Only Earth got it just right. Thanks to its modest greenhouse effect, its average surface temperature is fifteen degrees Celsius, thirty-three degrees warmer than it would be without an atmosphere. Everyone agrees that is a good thing. Over billions of years the amount of CO_2 in the atmosphere has fluctuated a lot, just as

Arrhenius and Högbom suspected, as volcanoes spewed more or less of it into the air. The underlying mechanism is more complicated than they could have known, and it didn't begin to be understood until the discovery in the 1960s that Earth's surface is broken into moving tectonic plates; basically the amount of CO_2 coming out of volcanoes depends on how much seafloor limestone is being carried into Earth's hot mantle at places where plates collide and one dives under the other. That process of carbonate recycling changes much too slowly to explain the waxing and waning of ice sheets over tens of thousands of years during glacial and interglacial periods. But a steady decrease in volcanic emissions over millions of years probably does explain why Earth has been cool enough for the past 2 million years to have ice sheets at all—in that respect Arrhenius was right. Ninety million years ago during the Cretaceous Period, when breadfruit trees and crocodile-like reptiles lived in the high Arctic and the seawater there may have been as warm as twenty degrees Celsius, or even 50 million years ago during the Eocene, when temperate forests stretched from pole to pole, the amount of CO_2 in the atmosphere must have been several times at least what it is now.

The Eocene was the time Arrhenius had in mind when he calculated the effects of tripling CO_2. Laboring away in Stockholm, where the mean annual temperature is less than seven degrees Celsius, a tropical hothouse climate did not sound so bad to him. "How little interest do we take in a barren island of the Arctic Circle, on which not a single plant will grow, compared to an island in the tropics which is teeming with life in its most wonderful variety!" he once wrote dreamily. Arrhenius was a genial bon vivant, an optimist even by the standards of his pre–World War I era. He first hatched his idea about the greenhouse effect and the ice ages in December of 1894 and started his actual calculations the following July—the same month his ravishing young wife, Sofia, deserted him, after a year of marriage, and started writing him letters about how blissfully happy she was without him. In November, Sofia, now living alone on an island off Stockholm, gave birth to their son, whom she did not allow Arrhenius to see. Svante soldiered on. By Christmas he had finished his pioneering greenhouse model, which is still cited more than a century later, and was complaining to a friend that he found it "unbelievable that so trifling a matter has cost me a full year." Maybe his wife's departure had been for him too a liberation; she was a mystical sort who had lately fallen in with theosophists and had more than once tried to convince her portly husband to purify his body by abstaining from smoke and drink. Certainly Arrhenius had a large appetite for work as well as pleasure, and an ability to look on the bright side of things.

The prospect that humans might be making the current "genial" period (as Arrhenius called an interglacial) a bit more genial was far from alarming him. He was the first to call attention to it, in a public lecture in January 1896. His friend Högbom had estimated that the 500 million tons of coal then being burned annually in factories and homes were raising the CO_2 level by around a tenth of a percent per year. A doubling of atmospheric carbon dioxide, Arrhenius told his audience, was thus thousands of years away, and our descendants then would thank us for the "warmer sky" and "less harsh environment than we were granted." If anything, he wrote later, they might regret the lost coal:

> We often hear lamentations that the coal stored up in the earth is wasted by the present generation without any thought of the future . . . We may find a kind of consolation in the consideration that here, as in every other case, there is good mixed with the evil. By the influence of the increasing percentage of carbonic acid in the atmosphere, we may hope to enjoy ages with more equable and better climates, especially as regards the colder regions of the earth, ages when the earth will bring forth much more abundant crops than at present, for the benefit of rapidly propagating mankind.

It is an argument that a few global-warming skeptics continue to make to this day, though as we shall see, the research does not really bear it out.

After Arrhenius, the next person to revive the greenhouse theory of climate change was a British steam engineer and amateur meteorologist named Guy Callendar. In a 1938 paper, he claimed that mankind had already added 150 billion tons of CO_2 to the atmosphere, and that the temperature was rising by .3 degree Celsius per century as a result. But he too said the increase was "likely to prove beneficial to mankind," mentioning the same reasons as Arrhenius and throwing in another: "The return of the deadly glaciers should be delayed indefinitely." Callendar's work also excited widespread indifference. Next, in a series of papers in 1956, the American physicist Gilbert Plass developed the idea much further and pegged the temperature rise at 1.1 degrees Celsius per century; in one paper he even allowed as how the warming "may be so large in several centuries that it will present a serious problem."

But the turning point, and the dawn of the modern era of greenhouse studies, came the next year, 1957. It was a paper by Roger Revelle and Hans Suess. Revelle was director of the Scripps Institution of Oceanography in La Jolla, California; Suess, a physicist, had joined him there in 1955. The two of them used radiocarbon data to estimate how long a molecule of CO_2 stays in the atmosphere before it dissolves in the sea—about ten years, they figured. That meant the ocean should rapidly take up any carbon dioxide we put in the atmosphere, as scientists who dismissed the idea of any significant human contribution to the greenhouse effect had always argued. On the other hand, Revelle knew from his own studies of the complex chemistry of seawater that the ocean might be spitting a lot of that CO_2 right back out again. He and Suess were torn; Callendar had thought all the CO_2 from fossil fuel emissions was accumulating in the atmosphere, and they knew that couldn't be right, but they couldn't tell what the right answer was.

Their graph projecting the growth of atmospheric CO_2 didn't even take into account that fossil fuel emissions were already growing exponentially, as nearby Los Angeles and other American cities and the whole global economy boomed after World War II. Looking to the northern horizon from La Jolla, when the Santa Ana wind is blowing from the land out to sea, you could, then as now, see the plume of Los Angeles smog stretching out over the Pacific. But the potential environmental impact of economic growth had not yet sunk in. Revelle and Suess assumed for the sake of their calculations that we would always be burning just the same amount of fossil fuels that we were in 1957—and still they couldn't say, because they didn't know how much of that CO_2 would go into the ocean, whether the concentration in the atmosphere would increase over the next century or so by 2 percent or by 10. In fact it has already increased by more than 20 percent, just in the half century since their paper, and it is thanks indirectly to them that we know that. Their most remarkable insight has been quoted so often it has become a kind of mantra of the global warming age:

> Thus human beings are now carrying out a large scale geophysical experiment of a kind that could not have happened in the past nor be reproduced in the future. Within a few centuries we are returning to the atmosphere and oceans the concentrated organic carbon stored in sedimentary rocks over hundreds of millions of years. This experiment, if adequately documented, may yield a far-reaching insight into the process determining weather and climate.

While they were still writing that, in August 1956, they had already hired a young scientist from Caltech named Charles David Keeling to do the documenting—to measure in an accurate way for the first time the amount of carbon dioxide in the atmosphere.

Broecker, working on his PhD at Lamont, absorbed in the Ice Age, read the Revelle and Suess paper, and he read Plass's papers. In an article published in 1957 in *Yale Scientific*, a student magazine, he and another Columbia graduate student named Bruno Giletti went so far as to point out that "the warmer climate that might result [from a CO_2 increase] could raise such problems as coastal flooding due to rise in sea level and increased aridity in certain areas." Fifty years later, those remain the two greatest dangers of global warming. But Broecker and Giletti were not really alarmed then, any more than Revelle and Suess were. The "geophysical experiment" that Revelle and Suess were talking about, whose initial result Keeling would soon be revealing to the world, was not a fraught thing; it was a good thing. Experiments are how scientists figure out how the world works. And at that time, in the 1950s, what drove all of them was not so much alarm as curiosity.

Hans Suess had scientific curiosity in his DNA—he was the son and grandson of eminent Austrian geologists, professors both—and he had strong ideas about what got in the way of such curiosity. On first meeting him people would sometimes get the impression he didn't do much; he spent a long time thinking before he did any kind of experiment. When colleagues would speak of the meetings they were organizing and the committees they were chairing, Suess would ask, "When do you have time to think?" Broecker was in his first year of graduate school when he met Suess, who was then still at the U.S. Geological Survey in Washington. Like Broecker and Kulp, Suess had learned radiocarbon dating from Willard Libby in Chicago, but he was doing a better job of it. The Lamont team was still doing it the Libby way: converting carbon from the sample they were dating to solid carbon black, which they mixed with water and spun onto the inside of a cylinder as a wet slurry. Before they shoved the cylinder into their Geiger counter, they first dried the carbon by blowing it with a hair dryer. As it turned out, whenever the U.S. government detonated an atomic bomb at the Nevada test site, the air in the Lamont lab, like the air over much of the United States, would contain randomly fluctuating amounts of radioactive strontium and cesium, which the hair dryer would then deposit on the carbon samples. That contamination was making the radiocarbon measurements hopelessly erratic. Broecker and Kulp went to Washington to learn how Suess was avoiding this infuriating problem.

Suess was avoiding it by extracting the carbon as acetylene gas instead of carbon black, thus eliminating the need for a slurry and a hair dryer. Acetylene, however, is an explosive gas, and Broecker did not use Suess's method for long—but he also learned a more lasting lesson that day. A few minutes after he and Kulp arrived in Suess's lab, Kulp excused himself to go to the men's room. An awkward silence ensued, as the twenty-three-year-old Broecker sat facing the forty-four-year-old Suess. Broecker never did find out why the older man chose that moment, so early in their aquaintance, to share with him the secret of scientific life, but share it Suess did.

"Young man," the Austrian said gravely, "a disaster happens to many, many of our best scientists. They become administrators. And the day they do that, they're lost to science. So you never want to become an administrator. You have to guard against that."

"But Dr. Suess," piped Broecker, "how do I do that?"

"Be a dynamic incompetent! Do at least three outrageous acts a year. Then no one will *want* you to be an administrator."

Words to live by, Broecker would later decide—but we digress. Around that time, before joining Revelle at Scripps, Suess had the idea of measuring the amount of radioactive carbon-14 in tree rings. Carbon-14 is produced in the atmosphere by cosmic rays, incorporated there into a small percentage of CO_2 molecules, then taken up by plants as they grow. By measuring the amount of carbon-14 in different rings from the same tree, Suess wanted to see if the amount in the atmosphere was changing over time—which was important to know if you wanted to use radiocarbon as a dating method. He found that since the nineteenth century it had decreased by several percent. Assuming that the supply of the carbon-14 hadn't changed—that the galaxy wasn't sending us fewer cosmic rays—that meant the atmospheric CO_2 pool was being diluted by a source free of radiocarbon. Fossil fuels contain no radiocarbon. They are hundreds of millions of years old, and the carbon-14 that was in the organisms that made them has long since decayed away. Suess calculated that a rise in atmospheric CO_2 of 20 percent or so due to fossil fuel burning could explain his results. But his estimate was highly uncertain. He and Revelle knew that. That's why they hired Dave Keeling.

Keeling had gotten a PhD in polymer chemistry at Northwestern—he grew up in the Chicago suburbs, during the same period as Broecker. But chemistry in general, and polymers in particular, never moved him much. By the time his dissertation made the cover of *Chemical and Engineering News*, he had already begun casting about for a different path in life. A chance encounter with a book called *Glacial Geology and the Pleistocene Epoch*, at a time when he did not yet know the word *Pleistocene*, had gotten him interested in geology. On receiving his degree he turned his back on job offers from chemical manufacturers back East, thus forsaking a career in plastics, and instead applied for work in geology departments west of the Rockies. His father had been raised in Montana. Keeling liked the outdoors. He wanted work that would take him there.

When he arrived at Caltech in Pasadena for his first job, his boss, a geochemist named Harrison Brown, was busy writing a dark little book called *The Challenge of Man's Future*. Brown had become doubtful of the future in part as a result of his work isolating plutonium for the Manhattan Project, but the greenhouse effect was not one of the challenges that interested him. An offhand remark of his nevertheless set Keeling on his career path. Brown suggested that the acidity of groundwater or a river would be determined by how much carbon dioxide was dissolved in it—the more CO_2, the more acid the water—and that in turn would reflect a chemical equilibrium between the water and the air. Keeling decided to find out if his boss was right by going out and measuring CO_2 in the water and the air.

He set out in May 1955. The rivers in Southern California were dry; he had to drive as far north as Big Sur to find one with water in it. That river flowed through a redwood forest, and to get an accurate fix on the CO_2 in the air there, Keeling felt he had to camp in the forest for days and measure CO_2 at night as well as in the daytime. At 2:30 a.m. on a clear, nearly moonless night, with starlight flooding down through the redwoods, Keeling stood on a footbridge over the Big Sur River, opened the stopcock on an evacuated five-liter glass flask, and let the cool, damp air rush into it, as it rushed too into his own young and eager lungs. In June, according to the data chart he later published, which reads strangely like a travel diary, he was on the shore of Lake Tenaya in Yosemite National Park; July found him on the other side of the Owens Valley, at fourteen thousand feet in the White Mountains; and in early September, while Broecker was tromping around Pyramid Lake with Phil Orr, Keeling was standing on another footbridge and then on a beach on the Olympic Peninsula in Washington, which was just about the only place he saw clouds in the sky he sampled that year. In 1956 he visited some of the same places again, as well as Organ Pipe Cactus National Monument in Arizona. On a trip to Washington, D.C., that spring to speak to the director of meteorological research at the U.S. Weather Bureau, he managed a side trip to Chincoteague and Assateague islands, off the eastern shore of Virginia. He collected air samples there too.

There was no compelling scientific reason for this gadding about. But this was why Keeling had chosen geochemistry over plastics—and afterward the results proved more compelling than he could have guessed. Back in the lab at Caltech, he passed the air from his flasks through a copper tube bathed in liquid nitrogen, which froze out both the carbon dioxide and the water vapor in the air. When he next allowed the tube to warm, but still kept it chilled by dry ice, the water stayed frozen, and what wafted out of the tube was the carbon dioxide gas, all of it and nothing but, from the original air sample. Keeling measured the pressure generated by that gas, and thus its concentration, by seeing how much it displaced a column of mercury in a manometer—a U-shape glass tube. He had designed it himself.

Keeling quickly found that Brown's conjecture had been wrong: river water was not in chemical equilibrium with the air above it, because the rivers were picking up lots of carbon from the rocks they flowed over. After he figured that out, he stopped sampling water and focused on the air. There was a sensible rhythm to what he found there: the carbon dioxide concentration was always lower during the day, when plants around his sampling sites took up CO_2 and used it in photosynthesis to make sugars. Then when the sun went down and photosynthesis stopped, the plants burned some of those sugars and respired CO_2, just as we do, raising the concentration in the atmosphere again. Keeling had measured the breath of those redwood trees at Big Sur.

But he had also found something much more surprising. The baseline concentration, when it reached its minimum in the late afternoon, was pretty much the same everywhere he went. The conventional wisdom had been that the carbon dioxide concentration varied from one place to the next, from as low as 150 parts per million to as high as 350. One famous meteorologist had even proposed that you could use CO_2 as a chemical tag to track air masses of different origin as they moved around the planet. But from Assateague to Big Sur to the Olympic Peninsula, near the heavy breathing of trees or far from it, Keeling consistently measured a concentration of around 315 to 320 parts per million. Random laboratory error could explain other researchers' highly variable numbers, but not Keeling's consistent ones. CO_2 seemed to be well mixed around the planet; the uniform background concentration was modified only slightly by local effects. Keeling's pleasure-seeking travel had produced his most important result.

What it meant is that you could hope to find out whether atmospheric carbon dioxide was increasing over time with measurements made in a few places, or even in only one. That's what Revelle and Suess hired Keeling to do. He moved to Scripps in the summer of 1956—relieved to reject the basement office in Washington that the Weather Bureau had offered him in favor of the sea breezes of Scripps, which is perched on a cliff above the Pacific in La Jolla. Keeling's first problem was to figure out a way to speed up and scale up his hands-on measurements; in his first year and a half he had made just a few hundred. But at the Applied Physics Corporation in Pasadena he had met an engineer who showed him a device that might do the job: an infrared gas analyzer. Originally built for military and industrial purposes, it used the principle of the greenhouse effect itself—the fact that certain compounds absorb infrared light—to measure their concentration in an air sample. The device wouldn't eliminate the need for Keeling's five-liter flasks and his manometer—they were by far the most accurate way of measuring CO_2, and they were needed to calibrate the infrared device—but it would allow him to measure CO_2 continuously at a given location.

In March 1958, Keeling began measurements at an observatory the Weather Bureau had established on the north slope of Mauna Loa in Hawaii. At an altitude of over eleven thousand feet, and in the middle of the Pacific, the observatory was away from large industrial sources. Yet in that first year Keeling worried that his data were seriously flawed. They fluctuated so widely—first rising to a peak of 315.1 parts per million in May, then falling continuously for six months, to a low of 310.6 in November. When they began to rise again in December, though, Keeling realized he had discovered something real—a seasonal cycle of CO_2 to go with the diurnal one. Just as plants draw CO_2 out of the atmosphere during the day and return it at night, they draw it out in spring and summer, during the growing season, and return it in winter. The atmospheric concentration of CO_2 thus peaks in May, just before the growing season begins. (It follows the northern seasons because there is more land and more vegetation in the northern hemisphere than in the southern.) On Mauna Loa, Keeling was no longer measuring the collective breath of a single stand of redwoods; it was the whole planet breathing this time.

It should have been regular, like a sine wave, with the CO_2 value beginning each annual cycle at the same point. It *was* regular like a sine wave—but it quickly became evident that each year's starting point was not the same as the last. "When I first published the data in 1960," Keeling recalled in an interview in 2005, "we had enough data to see it was higher the second year. And then it was higher the third year. And then it was higher the fourth year. Then we knew something was going on."

Year after year the pattern persisted; slowly the oscillating curve sloped upward, at first at the rate of less than a part per million a year. Slowly too the message that atmospheric carbon dioxide had reliably been measured at last, and that it really was increasing, seeped into the scientific and public consciousness. To some scientists listening to Keeling present his results in the 1960s, even to Broecker, his steadily rising CO_2 curve came to seem a little monotonous. Keeling faced determined pressure from government grant-givers to move on to something else, and to turn over the Mauna Loa operation to the weather service, on the theory that CO_2 monitoring had become as routine as reading a thermometer. The pressure persisted throughout Keeling's career. He never gave in to it. When he died in June 2005, while hiking on his ranch in Montana, his Scripps lab was still measuring CO_2 on Mauna Loa. The value he got for 2004 was 377.43 ppm, nearly 20 percent more than when he started.

"Keeling's a peculiar guy," Revelle told an interviewer in 1989. "He wants to measure CO_2 in his belly . . . And he wants to measure it with the greatest precision and the greatest accuracy he possibly can." That single-mindedness and passion for accuracy made him difficult to work with sometimes. And yet Keeling was anything but a narrow-minded nerd. He was just as passionately an

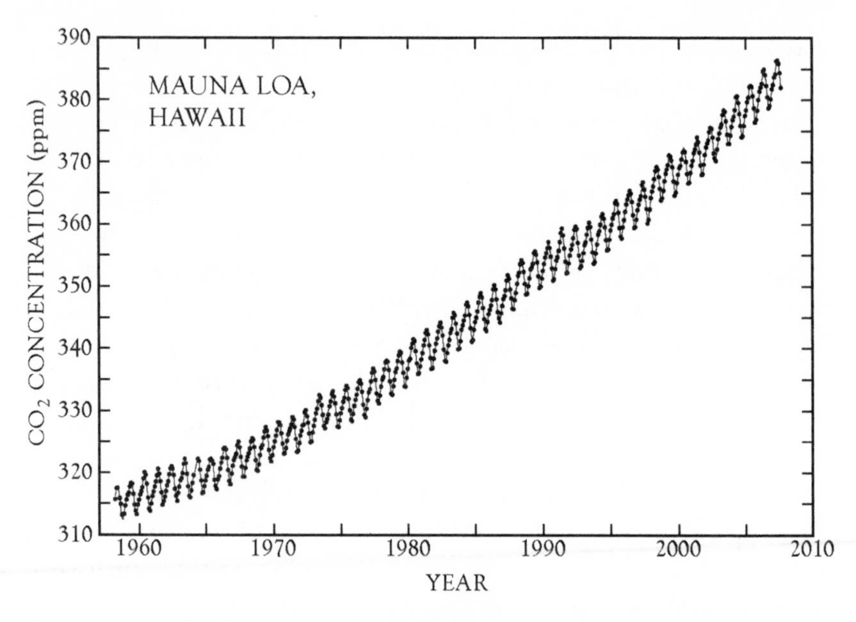

Since Dave Keeling began measuring CO_2 at Mauna Loa in 1958, the amount in the atmosphere has increased by more than 20 percent.

outdoorsman and a pianist—in his youth he gave paid recitals at women's clubs—as he was a laboratory scientist. A tall, handsome man, with a weathered face and a full head of gray hair, he spoke plainly, in slow, distinct, complete sentences, but with erudition: in addition to *Glacial Geology and the Pleistocene Epoch*, he had read Gibbon's *Decline and Fall of the Roman Empire* in its entirety, twice. When he delivered pessimistic opinions about the future of the American empire, or of humans on Earth more generally, he did so with a glimmer in his eye and an open, bemused look on his face. At heart he was perhaps not really a pessimist. His father, an advocate of banking reform in the years after the Great Depression, had turned Keeling off economics for life by administering a childhood overdose. But he had also passed on to his son a "faith that the world could be made better by devotion to just causes."

One man, even a quiet scientific man, can make a difference in the world: if Keeling had not been so devoted to measuring carbon dioxide, the debate on global warming would be even more mired in polemics than it is now. Instead, the "Keeling curve" of carbon dioxide at Mauna Loa has become one of the debate's few universally acknowledged truths. Even skeptics agree that carbon dioxide is a greenhouse gas; even skeptics agree, thanks to Keeling, that it is increasing; and even skeptics agree, thanks again in part to Keeling, that the increase is due to our burning fossil fuels. Building on the isotopic techniques pioneered by Suess, he was among the first to show that the CO_2 in the atmosphere bears the distinct chemical signature of fossil fuels. In 2002, Keeling received the National Medal of Science from President George W. Bush, the climate skeptic-in-chief—and the symbolism of that, to anyone who still believes in the power of science to affect everyone's view of reality, was heartening.

Of course, it does not really take isotope ratios to show the connection between burning fossil fuels and CO_2 accumulating in the atmosphere. From Arvid Högbom through Guy Callendar to Dave Keeling, and since the 1950s with the help of the United Nations, people have compiled increasingly accurate statistics on fossil fuel use—and have learned, among other things, just how wrong it was for Revelle and Suess to ignore in their projections the exponential growth that was happening all around them. "The consumption of fossil fuel has increased globally nearly three-fold since I began measuring CO_2 and almost six-fold over my lifetime," Keeling wrote in 1998, reminiscing on his career. On Mauna Loa he could track that growth precisely, year by year, with his upward-sloping CO_2 curve. But there was no place quite like his Southern California home to get a visceral feel for it.

> When my family first moved here over 40 years ago, we could stand on a vantage point above the Pacific Ocean a few miles north of the city center of San Diego and look eastward over an expanse of hills and distant mountains accessed by only a few country roads and inhabited mainly by farmers and wildlife. Revisiting this vantage point at night many times since, I have watched the number of lights steadily increase; lights from new homes, lights from commercial enterprises, lights from vehicles after an eight-lane highway was built

> just to the east 30 years ago. I have repeatedly asked myself how long these increases can go on.

They have already gone on longer, as far as climate is concerned, than, say, Guy Callendar might have thought—Callendar thought all the carbon dioxide emitted by burning fossil fuels stayed in the atmosphere. If that were the case, the concentration in the atmosphere would now be approaching 500 parts per million instead of 400; and the change in climate would already probably be such as to preclude all debate about what caused it. What Keeling's record has consistently showed, however, when compared with records of fossil fuel use, is that only half the carbon dioxide we are putting into the atmosphere is staying there—the rest must be going into either the ocean, as Revelle suggested, or into plants on land. Over the years both Broecker and Keeling himself have had a big hand in the search for this "missing carbon." But probably the man who has done the most convincing job of finding it is Keeling's son, Ralph.

SIX

Where the Carbon Goes

In November 1957, while Keeling was struggling to get the Mauna Loa station set up, a young Japanese named Taro Takahashi was standing on a dock in Piermont, New York, on the Hudson River, just a couple of miles north of the Lamont lab. He was about to board Lamont's research ship, a two-hundred-foot, three-masted schooner called *Vema*. Takahashi had gotten his PhD from Columbia the previous spring, like Broecker, and his path to that dock had begun shortly after graduation in the Columbia geology department—more precisely, in the men's room. (The role of men's rooms in scientific progress has too often been ignored by historians.) There the slight young man with the thick accent had found himself looking up one day at a tall and imposing Texan, who was busy at the next urinal. It was Maurice "Doc" Ewing, director of Lamont. Did Takahashi have a job lined up? Ewing asked. No, sir, he did not, said Takahashi. He had planned to be a mining engineer, like his father and grandfather before him in Japan, but there was a recession going on. Ewing offered him a job on the spot. Would Takahashi like to cruise the Atlantic to see if it was soaking up carbon dioxide—in particular the carbon dioxide from all the coal people were mining and burning?

Vema had been built as a pleasure yacht for E. F. Hutton, the Wall Street financier, and it had once had fancy plumbing and a grand piano. But under Ewing's leadership and Ewing's budget, conditions on board had changed. Ewing hired Takahashi, at $4,000 a year, under the auspices of the same government research program that had brought Keeling to Scripps, a sixty-seven-nation collaboration called the International Geophysical Year. Takahashi was told, however, to bring his own shower curtain—not for showers, which were in any case allowed only twice a week, because *Vema* lacked a freshwater generator, but to suspend above his bunk, so he would not get drenched at night by seawater leaking through the teak deck. Nor was that the young graduate's only reason to be apprehensive about his first job. Ewing was on the dock that day to see him and the rest of the party off, and Takahashi took the liberty of inquiring when he might expect to come home. Ewing pointed out that Charles Darwin had shipped out on the *Beagle* for five years. Takahashi would just have to wait for instructions to be cabled to one of his ports of call.

In the end he got the cable in Cape Town only ten months later. By then *Vema* had been all the way down the Atlantic to Antarctica and back, stopping every day and using its sails to hold station for a variety of measurements, including Takahashi's time-consuming water sampling. His most painstaking operation was the one he did for Broecker—collecting 450 liters of water from depths of a mile or three, using a barrel with spring-loaded lids on a long cable. On deck he would pour a bottle of sulfuric acid into the water, which converted all the carbon dissolved in it to carbon dioxide. Then he would bubble nitrogen through the water to strip out the carbon dioxide—all this on a small, rolling ship, all this so that Broecker would have enough carbon to radiocarbon-date the deep water, that is, to determine how much time had elapsed since the water had been at the sea surface and had absorbed carbon, radioactive or otherwise, from the atmosphere. Meanwhile Takahashi himself was monitoring that exchange directly. Using the same infrared gas analyzer that Keeling used, he measured pCO_2—the vapor pressure of CO_2—in both the surface water and in the atmosphere. Whichever reservoir had a lower pressure would be absorbing CO_2 from the other.

Little did Takahashi know as he left the land behind in that adventurous year that he was leaving his mining career behind too, as Keeling had forsaken plastics, and embarking as Keeling had on a measurement he would still be doing nearly half a century later—monitoring the breath of the ocean. For as Takahashi has found, the ocean breathes as the land does, seasonally but also regionally. In tropical oceans, where trade winds push the surface waters aside, deep water wells up to the surface, warms in the sun, and discharges its rich load of CO_2 into the atmosphere. Closer to the poles, and especially in the seas around Iceland and Greenland, surface waters cool, absorb CO_2 from the atmosphere, and sink into the deep.

Even before human beings came along, the world ocean was a patchwork of such CO_2 sources and sinks. Globally, though, the two were in rough equilibrium: as much carbon dioxide was bubbling out of the sea as was dissolving back into it from the atmosphere. Human beings have disrupted that balance. By extracting large quantities of fossil carbon from the land, and pumping it into the atmosphere, we are increasing the pressure of CO_2 there. Indirectly but inevitably, we are pressing CO_2 into the ocean. But how much of it and how fast? The accumulation of fossil carbon in the ocean cannot be measured as readily as Keeling measured the buildup in the atmosphere, because the ocean started out with fifty times more carbon. The change caused by fossil fuel burning is thus relatively small and hard to detect—nothing like the 20 percent increase in the atmosphere that Keeling documented over his career.

If the fossil CO_2 were just dissolving in surface water, the amount soaked up by the ocean would be insignificant as well as undetectable; the water would soon become saturated, and its CO_2 pressure would balance that in the atmosphere again. But the CO_2 doesn't just dissolve—it is converted into other forms of carbon and thereby removed from the surface water, making room for more to dissolve in from the air. It is removed by microscopic floating plants, the phytoplankton, which capture sunlight and convert CO_2 into carbohydrates, just as plants on land do, or into calcium carbonate for their shells. It is removed in even greater quantities by complex chemical reactions, in which a CO_2 molecule together with a water molecule (H_2O) and a carbonate ion (CO_3) form two ions of bicarbonate (HCO_3). (Ninety percent of the carbon in the ocean is in the form of bicarbonate.) Finally, some of all that carbon gets mixed deep into the ocean by winds and ocean currents, or in the case of plankton corpses and other dead organic matter, simply by sinking. Over time, the whole ocean thus participates in taking up carbon from the atmosphere.

By 1957, when Takahashi set out on *Vema*, Broecker's radiocarbon work had already provided the first good estimate of just how long it takes for carbon to spread through the whole ocean. After he and Kulp got back from their visit to Hans Suess's lab, they reformed the Lamont radiocarbon operation, so that it would no longer be vulnerable to contamination resulting from nuclear tests. Broecker dated samples of deep water from the Atlantic and found that it had been several hundred years since that cold, dense water had sunk from the surface near Iceland; for it to spread all the way around the world ocean, he later calculated, must take on the order of a millennium. Clearly then the pulse of fossil carbon we began releasing in earnest in the nineteenth century was still making its way into the deep ocean.

To estimate how fast, Broecker had the idea of using a different kind of radiocarbon—a kind of silver lining to those mushroom clouds in the Nevada desert. The H-bombs produced radiocarbon the same way cosmic rays did: neutrons from the explosions collided with nitrogen atoms in the atmosphere, dislodging a proton and transmuting the nitrogen into radioactive carbon-14. Between 1952 and 1963, when the United States and Russia agreed to ban atmospheric nuclear tests, the tests produced a huge but temporary spike in atmospheric radiocarbon—its level nearly doubled. Then it gradually declined, as the excess radiocarbon was taken up, in the form of radioactive CO_2, by the ocean and by plants on land. By tracking the spread of that radiocarbon spike, Broecker and others realized in the 1950s, you could track the flow of carbon; the bomb radiocarbon was like a dye injected into a river. As early as 1959, for instance, Broecker published a little note in *Science* reporting that the average carbon atom in human blood was replaced every six months.

In the ocean, bomb radiocarbon is taken up at the same rate as fossil fuel carbon; the difference is that, being radioactive, and having been released at a precise time, it is easily distinguishable from all the carbon that was already in the water. In the 1970s Broecker, Takahashi, and other scientists got a chance for the first time to measure radiocarbon and other constituents of seawater all over the world. As part of a $25 million government research program called GEOSECS—which stood for Geochemical Ocean Sections—they steamed up and down the Atlantic, the Indian, and the Pacific oceans, and even into the Mediterranean and the Red Sea. All over the world, they found that the bomb radiocarbon had been mixed to depths of at least a few hundred meters by winds and storms; in the stormy Southern Ocean it was more like a kilometer. In one place, the North Atlantic, it had penetrated four kilometers down to the ocean floor. It had been carried there by the cold, salty water that sinks from the surface north of Iceland, then cascades over sills east and west of the island, down into the Atlantic abyss.

From the GEOSECS data, Broecker estimated that about one-third of the fossil fuel CO_2 is being taken up by the ocean. With half staying in the air, according to Dave Keeling's measurements, that left a sixth being absorbed by plants on land. For a time in the 1980s this estimate was controversial. Some researchers thought that forests and other land plants were taking up far more. Some thought on the contrary that the destruction of forests, particularly tropical rain forests—a problem that first gained wide attention in the 1980s—was releasing huge amounts of CO_2 into the atmosphere, as much perhaps as fossil fuel burning itself. But that debate is pretty much over now. The consensus is that Broecker's estimate of the relative importance of the ocean and land "sinks" for fossil fuel CO_2 is pretty accurate. It has been confirmed notably by much more sophisticated measurements made by Keeling's son, Ralph, who also has a laboratory at the Scripps Institution of Oceanography.

Since the late 1980s, Ralph Keeling has been measuring the amount of oxygen in the atmosphere. He got the idea from his dad while he was still in college, majoring in physics. Whereas plants give off oxygen when they take up carbon dioxide, Dave Keeling explained, seawater does not. Burning fossil fuels consumes oxygen, like any fire, so the level in the atmosphere must be falling—but to the extent that the resulting CO_2 is taken up by plants on land,

and they are spitting out oxygen, the oxygen level should be falling slower than it otherwise would. Measuring that decrease and matching it with the CO_2 increase would thus be a clean way of finding out how much of our CO_2 is disappearing into the ocean sink and how much into the land. The oxygen measurement is much more difficult, though, for the same reason that directly measuring the carbon buildup in the ocean is difficult—so much oxygen is already in the air that any decrease has to be minute. Normal air is nearly 21 percent oxygen, or 210,000 parts per million; that's around 550 times more than the CO_2 concentration of 380 parts per million. Dave Keeling always wanted to make the oxygen measurement but never figured out how to do it. He did however pass on its importance to his son. The idea lodged in Ralph's mind at just the phase in life when, if he's lucky, a young man realizes that his old man's life has not been totally without purpose.

After Ralph decided to become an atmospheric scientist rather than a physicist and was looking for a research topic, his father's problem came back to him. His physics training helped him invent a novel solution. It is based on the way air refracts light—which is something that can be measured with great precision using a device called an interferometer. Oyxgen, Ralph Keeling realized, refracts ultraviolet light more strongly than does nitrogen, which makes up 78 percent of the atmosphere; but nitrogen refracts visible light more strongly than does oxygen. By measuring the refractive index of an air sample at two wavelengths, one in the visible range and one in the UV, Keeling gets a precise measurement of the ratio of oxygen to nitrogen in the sample. And since the nitrogen level isn't changing, he can thus monitor the fall in the oxygen level to within one part per million.

He has been doing just that for a decade and a half at some of the same stations where his father monitored carbon dioxide—on Mauna Loa, for instance, at the Scripps pier in La Jolla, at the South Pole, and on Ellesmere Island in the Arctic. At each station the CO_2 curve and the O_2 curve are perfect counterparts. As CO_2 decreases every year during the growing season, the growing plants release oxygen, increasing its level in the atmosphere. But as the

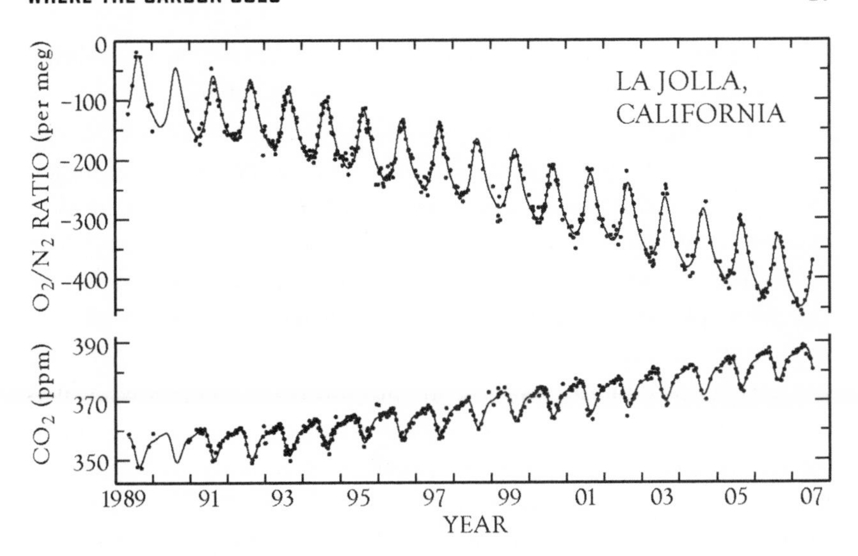

Ralph Keeling's measurements show that atmospheric oxygen (top curve) is slowly declining as CO_2 rises—and also that some fossil fuel CO_2 is being taken up by oxygen-releasing plants on land.

wiggling CO_2 curve slowly climbs from year to year, the wiggling O_2 curve slowly falls. Between 1989 and 2003, for example, the CO_2 level at La Jolla rose 20 parts per million, while the O_2 level fell by 49 parts per million.

If all the fossil fuel CO_2 produced during that period had stayed in the atmosphere, the concentration would have risen by twice as much, or 40 parts per million. And because it takes 1.4 molecules of O_2 to burn the average hydrocarbon molecule, the oxygen concentration would have fallen by 40 times 1.4, or 56 parts per million. (Burning gasoline or oil or natural gas uses more oxygen, because they contain hydrogen, which gets oxidized to H_2O; burning coal uses less, because it is almost pure carbon.) That the CO_2 concentration rose by only 20 parts per million indicates, as we've already said, that half of it was taken up by either the ocean or the land. That the O_2 concentration decreased by 49 parts per million instead of 56 indicates that land plants returned 7 ppm to the atmosphere. And by a fairly simple calculation we won't bore you with, it suggests that 35 percent of the fossil fuel CO_2 went into the ocean and 15 percent into the land.

It's worth understanding what kind of quantities we're talking about—it's worth looking at just a few more numbers. The average American emits about 20 tons of carbon dioxide every year; if you drive a car that gets twenty-six miles per gallon, which is much better than the national average of twenty-one, a pound of CO_2 comes out of your tailpipe every mile. Every year the United States as a whole pumps nearly 6 billion tons of CO_2 from fossil fuels into the atmosphere. Fortunately people in other developed countries use energy more sparingly—in Western Europe, for instance, the per capita consumption is typically around 60 percent of what it is in the United States. And just as fortunately for the planet, if not for the people themselves, many people in less developed countries are too poor to have access to fossil fuels at all. Their poverty brings the global emissions of CO_2 down to a bit more than four tons per capita, or 29 billion tons total. More than a fifth of that comes from the United States—which in one way is actually an improvement: a half century ago, before the rest of the world ramped up its energy consumption considerably, Americans produced nearly half the CO_2, though the volume then was of course relatively small.

Today the world produces more than three times as much CO_2 as it did in 1957. And Ralph Keeling's percentages suggest that each year we are pumping more than 10 billion tons of that CO_2 into the ocean via the atmosphere. It has occurred to many scientists and engineers that one way to combat global warming would be to cut out the middle step—to capture CO_2 at power-plant smokestacks and inject it directly into the deep ocean, without allowing it to pass through the atmosphere and change the climate first. These researchers tend to consider it a bitter irony that their idea has been vehemently opposed by environmentalists, who are concerned about the impact of all that CO_2 on ocean life. In the long run the CO_2 will get into the deep sea anyway. If we were to stop burning fossil fuels tomorrow, seawater would continue soaking the last century's worth of carbon dioxide out of the atmosphere, and over the next millennium the ocean currents would eventually spread it to every corner of the abyss, until only around 15 percent of what we had put in the atmosphere was left there.

Once we put it in the atmosphere, we cannot prevent it from seeping into the ocean. But of course we aren't going to stop emitting CO_2—for the foreseeable future, we are going to be emitting more every year.

In 1957 we didn't know what we were doing; now we do. Dave Keeling, like Roger Revelle and Hans Suess and Taro Takahashi and Wally Broecker, began studying CO_2 and climate out of sheer curiosity, and because he liked the work. By the time Ralph began following in his father's footsteps, that kind of pure and detached interest in our grand geophysical experiment, that kind of fun-loving innocence—though it would seem to go with Ralph's bangs and round tortoiseshell glasses and boyish grin—was no longer possible. Still, it was not a personal environmentalism that drove Ralph out of physics and into his father's line of research. Nor is environmentalism his prime motive today for continuing not only his own oxygen measurements but his father's measurements of CO_2.

"There was something about working on a problem that had very obvious historical relevance," he says. "To study something that couldn't be the same in two decades. There was a crying need for people to do what my father was doing, to study how the world was changing. If you don't make an invention in 1980, you can always make it later. But I was getting the satisfaction of knowing I was doing something that no one could go back and repeat. It was a way to feel like you made a difference. That was the appeal."

Thanks to the monitoring the Keelings have done over the past few decades, and thanks to the understanding of ocean chemistry gained through GEOSECS and its successor programs, we have some idea of what the future holds. For the next few decades or so, it seems likely that around a third of the carbon we put into the atmosphere each year will still get taken up by the ocean. After that things get more uncertain. Most of that uptake depends on the CO_2 reacting with carbonate in surface water, and the carbonate is slowly getting used up—converted into bicarbonate. As far as CO_2 goes, the oceans are slowly filling up. What's more, as they warm, their capacity to absorb more CO_2 from the atmosphere will decrease further, because the warmer surface waters will be less likely to sink and mix their load of CO_2 through the whole water column. Thus by the second half of this century, the oceans may be absorbing less CO_2 than they are now. "The oceans are the biggest natural sink we can count on," says Ralph Keeling, "but the sink is not big enough to do anything more than make a dent in the problem. It's a substantial dent, but the oceans are not going to save us."

In recent years, moreover, both Keelings noticed an interesting trend in their data. CO_2 and O_2 levels have always fluctuated with the seasons, but the amplitude of those seasonal cycles has increased—at La Jolla, for instance, the annual CO_2 excursions are 56 percent larger than when Dave Keeling started measuring them in the fifties. It is not clear why this is happening, but it must have something to do with photosynthesizing plants. In North America and Scandinavia, forests are regrowing on land once cleared for agriculture; and as the Arctic warms, faster than any other place on the planet, forests are also expanding north into regions once occupied by tundra. At the same time, because of the warming, the growing season is getting longer, especially in the far north.

All this suggests to Ralph Keeling an intriguing hypothesis for why the seasonal cycles of CO_2 and O_2 should be increasing. "The land is getting fuller and it's starting to hyperventilate," he says. "It's breathing hard." Which is an interesting observation, because some people had hoped that the land, not the sea, would save us from global warming.

Outside Durham, North Carolina, in a forest belonging to Duke University, a team of researchers has been trying for the past decade to find out whether that hope is well-founded. Similar experiments are under way in Nevada, Wisconsin, and Tennessee; they are all known by the acronym FACE, for Free-Air Carbon Enrichment. In the Duke Forest there are seven test plots, each of them a hundred-foot-wide circle containing 150 or so loblolly pines. Those trees are typically the first to return to abandoned farmland in the southeastern United States, and they're a fast-growing staple of tree plantations there. In the center of each circle is a square tower of hollow aluminum tubing, from which the Duke researchers make some of their measurements. As you ascend one of those towers, on a switchbacking series of sloping ladders, you eventually emerge from the shade into the bright sunlight at the treetops. The trees, planted as three-year-old seedlings in 1983, are now around fifty-five feet high. At the top of the tower you can reach out and touch the clusters of bright green needles that clearly demarcate last year's growth. And as the gentle breeze caresses your cheeks with 580 parts per million of CO_2—200 parts per million more than today and twice as much as before the industrial revolution began—you can ponder what a southeastern forest will look like in 2050 if we stick to "business as usual" and continue our relentless increase in fossil fuel consumption. So far, says Bill Schlesinger, the Duke geochemist who organized the experiment, it looks like the forest of the future "is one with pines growing slightly better, with a rank and lush understory of poison ivy."

Plant physiologists have known for decades that some plants grow better when there is more CO_2 in the air. To make proteins and fats, plants need nitrogen and phosphorus, which they get from the soil or the fertilizer we spread on it; but they need CO_2 to make everything. Commercial growers of tomatoes, say, routinely bleed extra CO_2 into the air of their greenhouses. And experiments in closed chambers with many other plants—even loblolly pine—had shown, before the FACE experiment began in 1996, that the plants grow faster when they have more of their most essential nutrient. That had led a few researchers to a happy thought: maybe in burning fossil fuels we are not so much polluting the planet as fertilizing it; maybe all those faster-growing trees will suck our excess CO_2 out of the air, thus protecting us from the worst of global warming, and at the same time giving us the benefit of all that extra wood. It was a happy thought indeed, and in principle it could even be true—but you can't find out in a greenhouse. Tomato growers and plant physiologists alike make sure the plants in their greenhouses don't run out of anything else they need, such as nitrogen or phosphorus and water, and they protect them from weeds and other pests. None of that happens in a natural forest. And so the question of how a natural forest will respond to enhanced CO_2 can't be settled in a greenhouse—it can only be answered in the forest itself. "Nature is not one plant in a pot, watered and sprayed and fertilized," says Schlesinger. "Nature is a mess. It's competition and poor soils and fungus and drought."

Hence the FACE experiment. Every day, an eighteen-wheel tank truck rumbles into the Duke Forest from a gas company in Virginia and off-loads cold liquid CO_2 into one of two 250-ton tanks at the edge of the site. From there the CO_2 flows through the vertical pipes and aluminum fins of a two-story-high heat exchanger, where it draws so much heat from the surrounding air that it becomes a gas—and makes the area around the heat exchanger, with its frost-covered conduits, a cool and pleasant place to stand on a summer day in North Carolina. The gaseous CO_2 then travels through white plastic pipes to a series of thirty-two towers spaced around the perimeter of each of the circular test plots. Through tiny, regularly spaced holes in those pipes, it leaks into the air around the growing pines. A computer in a little shed adjusts the flow, based on the measured CO_2 concentration at various points inside the circle—the goal being to keep it as close to 580 ppm as possible—and also on the wind direction: the CO_2 is released only on the upwind side of the circle, so that as little as possible will go to waste. Though the whole reason for the experiment is that the world can't stop producing mind-boggling quantities of CO_2 as waste, the gas company salesmen call it "product" when they take Schlesinger to lunch, and to deliver it in a nice, pure, and convenient form, they charge $65 a ton, about twenty-five cents a gallon. In all, the Duke experiment costs about $700,000 a year for CO_2 alone.

The scientists go to great lengths to track the product after it loses itself among the loblollies, and to find out how much of it is getting locked away in wood or soil. The floor of the forest is dotted with plastic laundry baskets that collect falling pine needles, cones, and seeds. Plastic pipes angled into the soil allow cameras to monitor the growth and disappearance of roots. Yellow tape delineates "no-step" zones, where researchers measure the growth of seedlings. (There are not many seedlings, because deer ignore the yellow tape.) And the most important measurement of all is the one made by the expandable steel bands strapped around nearly every tree. Like tape measures around human waistlines, they show how much the trees are growing in response to their high-carbon diet.

They are indeed growing faster. Three of the circles in the Duke Forest are control plots, equipped in the same way as the four others but with ambient air flowing out of the pipes instead of CO_2. After eight years of the experiment, the forest is growing around 15 percent faster in the test plots than in the controls. That forestwide average masks a wide range of effects on different plant species. Poison ivy has benefited most from the extra CO_2, growing 77 percent faster; the loblollies have increased their growth rate between 13 and 27 percent, depending on the tree and the year.

To grow faster the plants have to be drawing more carbon out of the atmosphere, but not all of it stays out of the atmosphere. Not all of it is going into fatter and taller trunks; around a third is going into needles and cones, which litter the forest floor. Advocates of the CO_2 fertilization effect have argued that in the greenhouse world huge quantities of carbon would end up getting locked into the humus of forest soils—that the fast-growing trees would essentially be pumping CO_2 into subterranean storage lockers. But Schlesinger and his team don't see that happening. The amount of litter getting converted to humus is no greater in the test plots than in the controls. Instead, most of it is getting decomposed by fungi and bacteria, which respire the carbon back to the atmosphere.

With the results of their experiment so far, the Duke researchers have been able to estimate just how much help we might expect from forests in combating global warming. If the world's forests were all to respond to increased CO_2 as the Duke Forest has, then by 2050 they would be drawing an extra 3.4 billion tons of CO_2 out of the air. That sounds like a lot—it would nearly double the current land uptake of CO_2. But the trouble is, under the business-as-usual scenario, our emissions will have more than doubled. If the whole world responds as the Duke Forest has, and if we keep on the growth path we are on now, the land in 2050 will be taking up a smaller percentage of our CO_2 emissions than it is now, and the amount of CO_2 in the atmosphere will be growing even faster than it is now.

And that, Schlesinger says, is an optimistic forecast. Loblolly pine is one of the fastest-growing trees on Earth, and the ones in the Duke Forest are in their young and vigorous prime. It thus seems unlikely that the world's forests will on average respond to increased CO_2 as well as the Duke Forest has in the first years of the experiment. In the last few years, Schlesinger and his colleagues have noticed that, just as they feared, the soils in their test plots are running short of nitrogen; to keep the trees growing to their full potential, they've started fertilizing the plots with nitrogen, which is not something one could envision doing planetwide. (In the most rosy scenarios put forward by advocates of the CO_2 fertilization effect, the extra nitrogen would come from nitrogen oxides emitted by automobile tailpipes and ammonia wafting off fertilized farm fields; unfortunately or fortunately, that pollution doesn't seem to be enough at Duke.) As the forest matures, Schlesinger expects its growth to level off and the amount of carbon it stores each year to decline. A recent report from a FACE site in Switzerland reinforces that concern. That study was conducted not in a young pine plantation but in a natural forest with a variety of mature deciduous trees. It found that increased CO_2 did not increase the trees' growth rate at all over four years; nor did it increase the amount of carbon stored in the trees and the forest soil.

There is no doubt, though, that some of our fossil fuel carbon is disappearing into the land right now—around 15 percent of it, according to Ralph Keeling's oxygen data. Where exactly is it going? Certainly not into tropical forests, which are being logged and burned and converted to farms and pastures at a rapid rate. Not only the burning but also the farming pumps CO_2 back into the atmosphere: when farmers till the soil, they break up the organic matter in humus and release carbon dioxide. Globally such changes in land use are thought to add more than 5.5 billion tons of CO_2 a year to the 29 billion that fossil fuel burning is putting into the air. Somewhere on the planet, then, a land sink is counterbalancing that huge deforestation source and, on top of that, manages to store 15 percent of the carbon from fossil fuels. Faster-growing trees may well be a part of that sink.

But most of it can probably be attributed to land that was once farmed reverting to forest, in such countries as the United States. When the Pilgrims arrived in 1620, virtually all the land east of the Mississippi was forested; by 1920 virtually all of that virgin forest had been cut down. Since then, as the center of American agriculture has shifted to the grasslands of the Midwest and farms in the East have been abandoned, the forests there have been expanding again. That has been happening at just the right time to offset part of our exploding CO_2 emissions. But the process can't go on forever, and unlike the world's hunger for energy, it is inherently limited in scale. At some point the land will indeed fill up. Schlesinger has calculated that if *all* the agricultural land in the United States were converted to forest, it would soak up about one-third of the current U.S. emissions of fossil fuel CO_2.

How much CO_2 the land will actually absorb in the future is even more uncertain than the future of the ocean sink, and not only because it depends so strongly on human actions. The FACE experiments are still going on, and in any case they are not the last word on the subject; further research may yet reveal that the CO_2 fertilization effect is larger than they have suggested so far. (The other experiments in the United States have observed effects comparable to those observed at Duke.)

On the other hand, global warming itself will likely increase CO_2 *emissions* from the land. Since the mid-1980s, for example, wildfires in the western United States have occurred four times as frequently and have burned more than six times as much area as they had in the previous fifteen years. In 2004 in Alaska alone, fires destroyed 6 million acres of forest, an area the size of Vermont. According to a recent analysis, climate change is the most likely cause for the increase in forest fires—western summers have lately tended to be hotter and drier, and the snows in the mountains have been melting sooner, increasing the length of the burning season. Even when there is no burning, as soils get warmer, they will tend to release more CO_2, because the warmth will accelerate the metabolism of microbes and fungi and cause them to break down the organic matter in humus more rapidly. An analysis of data from a National Soil Inventory in Britain has recently documented this effect; it showed that the carbon content of soils all over England

and Wales, in forests, bogs, fields, and pastures, has significantly declined in recent years. The decline has been enough to completely offset the progress Britain has made since 1990 in reducing its fossil fuel emissions.

The bottom line is that the land is not going to save us from the effects of our fossil fuel habit, any more than the ocean will. Over the next few decades we can be optimistic and assume that the two of them together will continue to take up around half our CO_2 emissions, simply because the processes involved change relatively slowly. We should count ourselves lucky for that. But we can't expect nature to solve a problem we created. We will have to do that ourselves somehow.

On November 3, 1957, a few days before Taro Takahashi shipped out on *Vema*, a floppy-eared mongrel dog named Laika became the first living creature to orbit Earth. She had been picked up a few months earlier on the streets of Moscow and rushed into service by Russian space officials, who were eager to repeat the international sensation they had created a month earlier with their first satellite, *Sputnik 1*. That launch had been greeted around the world by banner headlines. In the United States, NBC and CBS had broken into their radio and TV programs to broadcast live the satellite's eerie radio beep, which made some people feel as if the Russians were watching them. *Sputnik 1* was still in orbit (though its radio had died) when it was joined in space by *Sputnik 2* and its passenger. In the close quarters inside the capsule, Laika lasted only a few hours before succumbing to heat and stress. Though the spacecraft's radio had no microphone to transmit her last frantic barks, it continued to beep at Earth for a few weeks. Somewhere in the middle of the Atlantic, sitting at *Vema*'s own radio set, Takahashi listened, as people all over the planet were listening, to those cricketlike chirps, one every three-tenths of a second. To a scientist it was an exciting, not an alarming, signal; it was the dawn of a new age.

Within little more than a decade, men had followed dogs into space, and the astronauts in *Apollo 8* were orbiting the moon.

When former U.S. vice president Al Gore gives his public lecture on climate these days, the one featured in the movie *An Inconvenient Truth*, the first slide he invariably shows is a picture taken from *Apollo 8* on Christmas Eve 1968. As the spacecraft came out from behind the far side of the moon and regained radio and visual contact with home, one of the astronauts, Bill Anders, photographed the blue-and-white Earth rising over the lunar horizon. That image, Gore points out, served as a galvanizing icon for the nascent environmental movement. Within two years activists had organized the first Earth Day, the U.S. government had created the Environmental Protection Agency, and Congress had passed a new and stronger Clean Air Act, as a direct result of which the air in American cities is much cleaner now than it was in the more smog-ridden fifties and sixties.

As we were gaining the ability to see the planet whole, we were also, more gradually, becoming aware of our ability to change the whole planet, and not just to sully isolated parts of it. Nineteen fifty-seven, the International Geophysical Year, was a milestone for both developments. The *Sputnik* satellites were the Soviet Union's signature contribution to the IGY, and under the same program Takahashi and Dave Keeling were beginning to take the measure of our "grand geophysical experiment" with carbon dioxide. Now we are half a century further along in that experiment, and to some extent the outcome has become clear. Carbon dioxide is accumulating in the atmosphere in direct proportion to our use of fossil fuels, which is to say at an accelerating rate, and the planet is getting warmer. When Keeling first noted the increase in CO_2, it was less than a part per million a year; now it is around two parts per million per year. In another fifty years, if we stick to business as usual, it will be more like eight parts per million a year, and we will be on our way to tripling the atmospheric concentration (compared to the preindustrial level) by the end of this century. Computer models suggest that a doubling of CO_2 would warm the planet by around three degrees Celsius on average, and a tripling would warm it by around five degrees Celsius, or nine degrees Fahrenheit. Interestingly enough, those numbers are close to the ones Svante Arrhenius got more than a century ago with pencil and paper, and knowing a lot less about how climate worked—which probably just means he got lucky.

The temperature rise could be less—the models produce a range of estimates—but it could also be more. There is in fact good reason to worry that it could be more. In general, the computer models have trouble reproducing some of the larger climate fluctuations that are now known to have happened in the past. A good example is the period that occurred 55 million years ago, at the end of the Paleocene period and the beginning of the Eocene—the period when, as Arrhenius noted, fossil evidence showed that the Arctic in particular was extremely warm. In the summer of 2004 a team of European and American scientists on a flotilla of three icebreakers plowed into the Arctic Ocean and retrieved the first deep-sediment core from the seafloor. The fourteen-hundred-foot-long cylinder of mud went as far back as the Paleocene/Eocene boundary. In that mud the researchers found the best evidence yet of just how warm the Arctic was then. The water temperature, they concluded, was seventy-three degrees Fahrenheit, forty-four degrees warmer than today—and eighteen degrees warmer than any model had predicted. Apparently the models, the same models we use to forecast the future of climate, are missing something—some feedback mechanism that amplifies greenhouse warming in the real climate, but that no scientist has yet been able to figure out.

The Eocene warm period is also a good illustration of the long-term cycle that regulates atmospheric CO_2—and perhaps of what the future might look like, millions of years from now, if human beings are not around. Whatever the missing feedback, the ultimate cause of the Eocene warm period was almost certainly, as Arrhenius and Arvid Högbom suspected long ago, copious CO_2 spewing from volcanoes. At the time, the tectonic plates carrying Africa and India were colliding with Eurasia, closing a shallow seaway called the Tethys, and plunging the limestones on the floor of that sea back into Earth's hot mantle. The heat dissociated the calcium carbonate in the limestone into calcium oxide and carbon dioxide, and volcanoes pumped the CO_2 back into the atmosphere. The beautiful thing is what happened next. In the hot climate generated by the high CO_2, more calcium oxide leached from soils on land and flowed through rivers to the sea. There shell-forming organisms combined it with CO_2 to make calcium carbonate again—and over millions of years, as those creatures died and sank to the seafloor, they pumped carbon dioxide back out of the atmosphere into thick beds of limestone.

Slowly the Eocene greenhouse evolved into an ice house. By 45 million years ago, judging from evidence found in the Arctic sediment core, the first icebergs were drifting in the Arctic Ocean. By 14 million years ago that ocean was beginning to be covered by ice, at the same time as a continent-covering ice sheet was growing at the other pole, on Antarctica. And by 2 million years ago Earth had cooled enough to where the slight fluctuations in sunlight arising from the Milanković cycles could start triggering an oscillation, every hundred thousand years, between deep glacial periods and warm interglacials. In the natural course of things, a glacial period would lie in our future, perhaps ten thousand or twenty thousand years from now, and then in another eighty thousand years or so another interglacial, and so on. Ultimately some renewed fluke of the tectonic cycle would once again overpower the orbital ones and return our climate to a hothouse like the Eocene.

There is beauty in these stately cycles, and what might seem like a comforting invincibility. No matter what we do, it might seem, we cannot disrupt them—cannot shake the rock-solid balance of nature. That is the conviction that previous generations of earthlings, who had not seen with their own eyes the finiteness of their planet, grew up with. But what we have learned from the space age and from the past few decades of environmental research is that nature is in balance only on time scales we don't care much about, only in the long run, in which we and our civilizations too will be dead. Over geologic time the natural carbon cycle has saved our planet from the fates of both Venus and Mars and kept it habitable. But it cannot protect our society from a grand geophysical experiment in which, in decades, we are returning to the atmosphere an amount of carbon that it took hundreds of millions of years to lock away in rock.

Worse, climate has turned out to be extremely sensitive to experiments that nature itself has done. That is another thing the computer models do not reproduce well—the pace at which climate has sometimes changed in the past. It has sometimes changed fast. The last ice age, for instance, was not a period of unbroken cold. Two-mile-long cores extracted from the Greenland ice sheet show instead that on repeated occasions the temperature there jumped out of ice-age into intermediate conditions, a change in Greenland of ten degrees Celsius or eighteen degrees Fahrenheit. The warming sometimes happened in a decade, then persisted for centuries. An eighteen-degree change in mean annual temperature is huge. It is the difference between the climate of Burlington, Vermont, and Birmingham, Alabama.

Every now and then, it seems, nature has decided to give a good swift kick to the climate beast. And the beast has responded as beasts will—violently and a little unpredictably. Computer models attempt to predict its future behavior by applying the laws of physics to the atmosphere, ocean, and land, and that is certainly a valid approach. But studying how the beast has responded in the past under stress is another way to prepare ourselves for what might happen as we take a whack at it ourselves. That's the idea that has obsessed Broecker for the past twenty-five years, and with each passing year it has come to seem more urgent.

SEVEN

A Conveyor Belt in the Ocean

It may have been in Bern, Switzerland, or in Evanston, Illinois. The year was 1984 if it wasn't 1983. And the man giving the lecture was either Hans Oeschger, the noted ice-core specialist at the University of Bern, or his colleague Bernhard Stauffer. Broecker's memory is a little fuzzy on those details. But it isn't the details that matter so much, it's the idea, and Broecker remembers vividly how he first got the idea that a sudden halt of the thermohaline circulation of the ocean—or the conveyor belt, as he later dubbed it—might cause a large and abrupt change in climate. It was listening to one of the Bern scientists describing their analysis of an ice core from Greenland—more specifically, of the bubbles of ancient air that had been trapped in the ice like time capsules for tens of thousands of years.

You could see them with the naked eye. As snow piles up on Greenland at the rate of two feet a year, the snow from previous years is buried and compressed, and gradually, over a century or so, most of the air is squeezed out of the spaces between the grains—most, but not all. By a depth of around 250 feet, the pressure has pinched shut the last tortuous conduits back to the surface, but tiny pockets of air are still locked in what is now solid ice. Ice from summer snow—it snows all year long in the interior of Greenland—contains far more bubbles, because the flakes are larger and fluffier than those of winter snow, and the resulting ice more porous. The summer bubbles form clouds of white in the gray ice of a core; and bands of those bubbles recur at regular intervals, like tree rings, marking the succession of years—tens of thousands of years. When the first deep ice core was extracted in the 1960s at an American military base in northern Greenland, a Danish researcher, Willi Dansgaard, who made isotopic measurements of the ice itself, not the bubbles, called his pioneering paper "One Thousand Centuries of Climatic Record from Camp Century on the Greenland Ice Sheet." Which just goes to show, as we've noted before, that you can find lyricism even in oxygen isotopes.

The oxygen isotopes that Cesare Emiliani had measured in seafloor sediment cores had revealed not so much the temperature of the seawater, as how much water had been removed from the sea during the last ice age and locked up in continental ice sheets. The isotopes in the Camp Century ice core, however, did really reflect the local temperature in Greenland—the temperature of the air, long ago, that made the snow that became a particular layer of ice. Dansgaard's logic, which he derived from widespread measurements of today's precipitation, went like this. As air masses move from the tropics toward the poles, they cool, and the water vapor they picked up from the warm ocean gradually condenses and rains or snows out of them. Each time that happens, the clouds get isotopically lighter: poorer in heavy oxygen-18, which falls out of a cloud more readily. By the time an air mass reaches Greenland and drops what's left of its water vapor as snow, it is very cold and very dry—and the colder and drier it is, the more depleted it is of oxygen-18. In the last ice age, Dansgaard's Camp Century measurements showed, the air over Greenland was far colder and drier than it is today. Even more interesting, it was not uniformly cold and dry for all those tens of millennia. There were spikes and troughs in the oxygen-isotope curve, and thus in the temperature in Greenland.

No one paid much attention to those at first—the Camp Century core was the only one of its kind, and so it was easy to interpret the spikes as noise in the data. Dansgaard himself was more concerned to argue that there were more regular rhythms to the temperature fluctuations, cycles of 80 and 180 years that he thought might correlate with cycles in the output of the sun. Those supposed cycles impressed Broecker too. His calculation that they had been cooling Earth and masking the greenhouse effect in the 1950s and 1960s, but would soon switch sign and start amplifying it, was what led him to sound the alarm in 1975 about global warming. But now as he sat listening to the scientist from Bern—it was probably 1983 and probably Oeschger—he was hearing a message that was potentially more alarming still.

Oeschger was presenting new results from a second Greenland ice core, from a place called Dye 3 in southern Greenland, about 120 miles in from the east coast. He and Dansgaard had been directly involved in extracting the core this time, with an American scientist named Chester Langway, and with the invaluable logistical support of the U.S. military—Dye 3 was one of a chain of early-warning radar stations. In a six-story golf-ball-like structure perched on steel stilts ten feet above the drifting snow, bored airmen watched radar screens for Russian bombers, rarely if ever venturing outside their lavishly equipped, overheated quarters. Meanwhile the scientists set up their core-analysis lab in a trench dug into the ice. The drill itself was in a shed next door. When they began drilling in 1979, the Danish, Swiss, and American flags flapped proudly from the rooftop. When they finished drilling in 1981, the flags still flapped, but the roof itself was buried in the snow.

Like the Camp Century core, the Dye 3 core penetrated right through the ice to bedrock—a distance in this case of a mile and a quarter. Dye 3 was nine hundred miles away from Camp Century, and yet when Dansgaard had plotted its oxygen-isotope curve, he had found the same violent shifts during the last ice age. In less than a millennium, the temperature seemed to climb halfway out of its deep glacial cellar, only to plunge back just as abruptly several centuries later. If the shifts occurred in both southern and northern Greenland, they were not likely to be just noise. They must reflect rapid changes in climate that had really taken place, in Greenland at least.

And in his lecture, Oeschger seemed to be providing a clue to what caused the changes. Researchers had been trying for decades to measure reliably the CO_2 in ice-core bubbles, and just a year earlier, in 1982, the Bern group had finally succeeded. Using a "dry-extraction" method—a device that crushed a tiny piece of ice between interdigitating sets of needles—they had obtained a result that has since been verified many times over: the carbon dioxide concentration in the ice age was two-thirds what it was before the industrial revolution began, and around half what it is today. But Oeschger went further that day. The CO_2 level in the bubbles, he said, seemed to have undergone some of the same sudden shifts as the oxygen isotopes, and at the same times. Dansgaard had suggested his oscillations might have resulted from "shifts between two different quasi-stationary modes of atmospheric circulation." Oeschger pointed instead at the ocean: as the storehouse of the planet's carbon, only it could explain rapid fluctuations in CO_2.

That's when Broecker had Broecker's Big Idea: the idea that ocean currents might rapidly change climate by switching on and off. And just as his 1975 warning that Earth was going to warm has proved true, even though Willi Dansgaard's 80- and 180-year climate cycles have never been seen again, so too has Broecker's idea survived and prospered and become a paradigm for an alarming new kind of climate mechanism—even though the rapid CO_2 fluctuations that Hans Oeschger was describing that day in Bern, or wherever it was, have long since been revealed to be an illusion.

In 1983, Broecker was feeling intimately connected to the ocean and its currents. He had spent a good part of the 1970s at sea with the GEOSECS expeditions. You probably have to have been on such an expedition—oceanographers call them cruises—to appreciate not only the tedium and frustration, but also the intense joy it can afford when it goes well, when the distance from land and its daily pressures leads not so much to a feeling of isolation but of relief, to a coming together of the small shipboard company behind some common purpose: Broecker now thinks of GEOSECS as the Camelot of his career. Oceanographic knights-errant sometimes get to have fun too. Postcards from Broecker's life in the 1970s would include scenes of his now thoroughly middle-aged conscience triumphing over his inner boy—in Reykjavik in 1972, for instance, when, before embarking as chief scientist on a GEOSECS leg that would take him straight down the Atlantic to Barbados on the RV *Knorr*, he resisted the temptation to steal a pawn from the chessboard after the last game of the Fischer-Spassky match. He regrets his pusillanimity now; the bottom of one of the giant barrels used for radiocarbon sampling would have been the perfect place to hide the pawn. And in other situations, more often than not, the inner boy won.

Steaming up and down and across the various oceans in straight lines, stopping at regular intervals to measure the same chemical properties over and over, does not sound like exciting, boyish fun. But Broecker and his colleagues were convinced, and history has proved them right, that they needed a basic matrix of data if they were ever to understand the chemistry of the ocean—and in particular to use that chemistry to map the currents. Messages in bottles or drifting buoys can map the wind-driven currents on the sea surface, but not the hidden ones in the deep. Analyzing the GEOSECS results, Broecker discovered that the total amount of phosphate and oxygen in a water mass didn't change much as that water moved through the abyss; you could use that chemical tag to distinguish, say, North Atlantic Deep Water, which sinks from the surface into the abyss around Iceland, from Antarctic Bottom Water, which forms in the Weddell Sea. You could even determine in what proportion the two were mixing at a given place. Meanwhile the radiocarbon measurements told you how old the deep water was. That showed how fast it was spreading from its place of origin.

With those two markers, Broecker could map the big picture of ocean circulation. He was of course not the first to ply those waters. The German ship *Meteor* had crisscrossed the Atlantic on latitude lines in the 1920s, measuring the temperature and salinity of the water from the surface to the bottom; and the great German oceanographer Georg Wüst, who would decades later surprise Broecker and John Imbrie in their shorts and academic robes, had used the *Meteor* data to identify the different water masses of the Atlantic, layered one on top of the other like strata of sediment. The differences in heat and salt content, Wüst and other oceanographers knew, give the water masses different densities, and the density differences drive them down into the abyss and through the ocean's interior and back to the surface. At the surface, this "thermohaline" circulation—from the Greek words for heat and salt—

collaborates with a quite distinct driving mechanism, the winds, to push water around in currents such as the Gulf Stream. Those are the ocean currents that have for centuries been familiar to mariners.

The driving mechanism of the thermohaline circulation is simple: the coldest, densest waters sink to the bottom of the ocean and spread out there. As they spread through the ocean, they are gradually warmed by solar heat filtering down from the surface. The warmed water slowly rises, and colder water coming in behind shoves underneath it. The Atlantic is like a bathtub with a spigot at each end, labeled North Atlantic Deep Water (NADW) and Antarctic Bottom Water (AABW), each of them open wide, always filling up the basin. The Pacific is different. At its northern end it has water that is just as cold as high-latitude water in the Atlantic—but the water is not as salty. So it's not heavy enough to sink all the way to the bottom; it stops at a depth of less than a thousand feet. There, as Broecker showed with the GEOSECS data, it floats on a mixture of NADW and AABW that has flowed in from the Antarctic and fills the abyss of both the Pacific and Indian oceans. In today's world, as far as the flow of water goes, the Atlantic is the dominant ocean.

Why is that? To answer that question, Broecker decided, you have to figure out why a ton of Atlantic water contains anywhere from two to four pounds more salt than a ton of Pacific water. Broecker came up with a theory. Seawater gets saltier through evaporation—the salt stays behind as water vapor rises into the atmosphere. Water evaporates from the ocean all over the world, then returns to the ocean through rain or rivers—but not to the same place, because winds move the vapor around. A careful look at a map of the world's winds and mountains showed Broecker that the Atlantic and the Pacific are not at all in the same situation in this hydrological cycle.

In the temperate latitudes of both hemispheres, the dominant winds flow from west to east. Winds flowing off the Pacific promptly smack into the Cordillera, the mountain range that runs the length of the Americas just in from the west coast. Cooling as they climb the western slopes of the mountains, they drop almost their entire load of water—that's why Pyramid Lake and the rest of Nevada, which lie in the eastern shadow of the Sierra Nevada, are in a desert. Winds flowing east off the Atlantic over Europe, however, encounter no such barrier. It's a long way across Asia, to be sure, but some Atlantic water vapor makes it all the way to the Pacific (after several more cycles of evaporation and precipitation). Meanwhile, in the tropics, where the trade winds flow from east to west, the situation is reversed. Winds flowing off the Indian Ocean hit the mountains of east Africa, and most of their water vapor never makes it to the Atlantic. But Atlantic water vapor sneaks through the gap in the Cordillera, over the Isthmus of Panama, and rains directly into the Pacific.

The result is that the Atlantic is always exporting water vapor to the Pacific—a large amount of water, an amount roughly equivalent to the flow of the Amazon River. It's as if every year the top three or four inches of ocean—except for the salt—were lifted off the Atlantic, moved through the atmosphere, and dropped onto the Pacific. If that were all that were going on, the Atlantic would be getting steadily saltier—but it is not. As the winds carry freshwater out of it at the top, Broecker realized, the thermohaline circulation is carrying salt out at the bottom. The deep current carries Atlantic salt around the tip of Africa and through the Southern Ocean, all the way to the Pacific. There it is reunited, after centuries of separation, with the Atlantic water vapor that has traveled through the atmosphere. Together again at last, they return home to the North Atlantic through the ocean's surface currents. Thus the cycle begins again.

Clearly the term *conveyor belt* doesn't capture the full romance of this globe-spanning affair. But that's the one Broecker came up with. And in the late 1980s his diagram of it, or rather the one an artist for *Natural History* magazine prepared according to Broecker's instructions, galvanized climate research. In its iconic simplicity—it basically depicted the circulation of the world ocean as a single loop, consisting of a broad blue band representing the cold, salty current out of the deep Atlantic and a red band to represent the return flow of warmer water at the surface—it completely ignored the innumerable real-world complexities of the circulation, the eddies

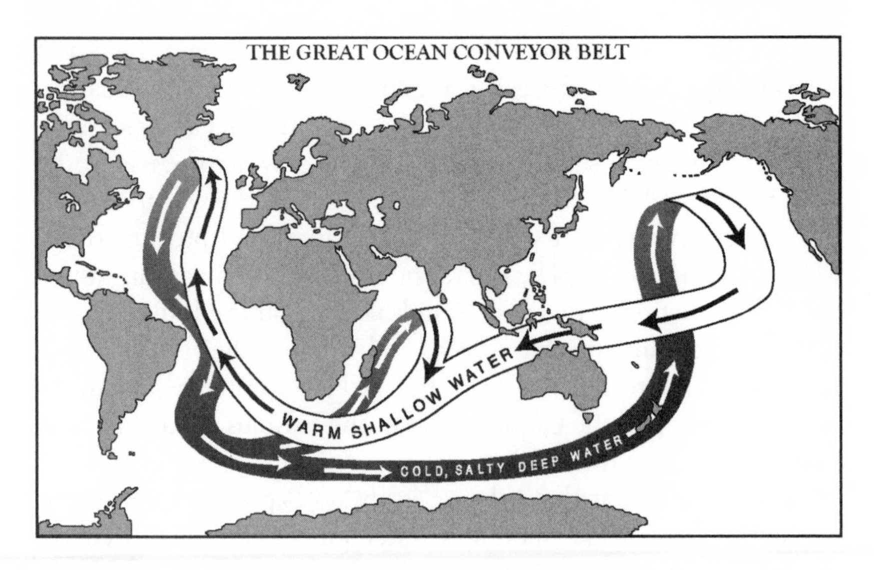

Broecker's "conveyor belt," an idealized version of the ocean's thermohaline circulation, transports heat into and salt out of the North Atlantic.

and gyres and countercurrents. It did that to focus on function—because in addition to conveying salt out of the North Atlantic, the conveyor belt conveys tropical heat back north. The winds that sweep over the sea surface around Iceland, carrying water vapor east, also absorb that heat and carry it to Europe. Greenland and the east coast of North America are heated to a lesser extent. If the conveyor belt were ever switched off, Broecker realized as he listened to Oeschger, the climate of the region would feel a deep and sudden chill. (Indeed, it has since been realized that when this happened during the Ice Age, sea ice covered the North Atlantic.)

What had happened to Broecker in that moment was a confluence of the two great currents of his research career: climate and the ocean. He himself was at a crossroads in his life. In 1980 he had been diagnosed with lymphoma and had twice undergone surgery to his jaw to remove the tumor. On New Year's Day 1981 he went to his doctor's home in New Jersey, where, amid the empty bottles and other debris of the previous night's party, the doctor removed the most recent stitches, because Broecker was leaving the next day on sabbatical. He spent most of 1981 in Heidelberg with Grace—their six kids were now grown—getting radiation therapy, but also writing a large book on the GEOSECS program, *Tracers in the Sea*. (It is long since out of print, but copies of it still sell for hundreds of dollars on eBay.) That fall he returned to New York, watched large bone tumors sprout on his forehead and the back of his skull, and started chemotherapy. By April 1982, just around the time the Bern group was publishing its first reliable data on ice-age CO_2, thereby drawing Broecker's attention away from the obscure chemistry and motion of abyssal waters, his cancer was gone. While undergoing chemotherapy he had turned fifty.

He had had good reason to believe his death might be imminent. Such knowledge affects different people in different ways. Broecker took refuge from it in work—but then, work had always been a refuge. Gloomy forebodings, he thinks, help explain why the publisher's fussy editing of *Tracers* got him mad enough to stomp off and publish the thing himself, thus embarking on a new sideline he pursued with subsequent books. But he sees no direct connection (though he agrees it makes a good story) between his near-death experience and his embarking on a new line of research, one that is of large public concern—he had always thought research should be useful as well as fun. In any case, since hearing that lecture of Oeschger's, Broecker has pursued almost obsessively the idea that climate has changed abruptly in the past. The concern is that it might do so again.

One result of his obsession is that the formerly obscure term *Younger Dryas* now crops up with great frequency in the scientific literature. *Dryas octopetala* is a woody, ground-hugging plant, a kind of rose, with cute little flowers, cream-colored and eight-petaled. It clings to rock and gravel in high mountain ranges in Europe and North America; it also flourishes in the northern tundra. It likes the cold, in other words, and it is a hardy colonizer. When the climate chills, it spreads down out of the mountains and south with the tundra. Decades before deep-sea foraminifera and oxygen isotopes were found to be tracers of past climate fluctuations, European botanists began using fossil *Dryas* plants for the same purpose. Analyzing cores from peat bogs, in areas far too warm today for the little flower to survive, they found layers filled with bits of *Dryas*—and concluded that those sediment layers had been deposited at a time when the climate was much colder than today.

That research showed, long before Dansgaard and Oeschger had seen their wiggles in the Greenland ice cores, that the end of the last ice age was fitful. It was not a long, steady warming accompanied by a steady retreat of the glaciers. After the Last Glacial Maximum, three resurgences of cold and ice were visible in the bog sediments—layers of gray silt and sand, littered with *Dryas* remains, sandwiched between the brown soils of the warmer periods. Of those cold periods, called the Oldest, the Older, and the Younger Dryas, the last was the most brutal. It started around 12,900 years ago, when temperatures had climbed almost to their levels today, and it lasted until around 11,500 years ago. Then it ended just as suddenly. Only then did the long, tranquil warmth of the Holocene begin; only then did the rise of human civilization begin.

The Younger Dryas showed up too in the Greenland oxygen isotopes—it was the most recent and one of the most prominent of Dansgaard's climate jumps. And by the time Broecker was listening to Oeschger's lecture, it had been documented in the sea as well as on land. Two Lamont colleagues of Broecker's, William Ruddiman and Andrew McIntyre, had seen it in a sediment core in the North Atlantic; during the Younger Dryas, a species of foraminiferan that now lives only in the Arctic had spread as far south as Britain. What that indicated, Ruddiman and McIntyre had decided, was that polar waters had surged south, diverting the Gulf Stream south too. Swinging like a door on its hinge, with the hinge at Cape Hatteras, the Gulf Stream had flowed east toward Spain rather than northeast toward Britain. It was the same course it had taken, Ruddiman and McIntyre said, at the height of the Ice Age.

To Broecker, looking for an explanation for the Greenland oscillations, that shift in surface currents looked like a symptom of something larger: a shutdown of the conveyor belt. What might have caused that? Ever the intellectual magpie, Broecker picked up an idea first put forward by an oceanographer named Claes Rooth at the University of Miami. If the engine of the conveyor was the sinking of cold, salty water in the North Atlantic, one way to jam it would be to make that water fresher; it would thus become too light to sink to the seafloor. At the end of the last ice age, as it happened, an enormous source of freshwater was right at hand—the Laurentide ice sheet. It covered most of Canada and a lot of the northern United States. Meltwater from its southern fringe accumulated in a giant glacial lake, named Lake Agassiz, which occupied a basin created by the weight of the ice; it was north and west of the present Great Lakes but larger than all of them combined. To the west and south, Lake Agassiz was dammed by large moraines; to the north and east it was bounded by the ice sheet itself. Its only outlet was to the south, via the Mississippi Valley, which carried torrents of meltwater to the Gulf of Mexico. But as the planet warmed and the ice sheet receded, a new eastern outlet eventually opened—rather abruptly, in Broecker's scenario. Ten million cubic kilometers of freshwater rushed through the Great Lakes Basin and down the St. Lawrence River to the Atlantic, spreading a buoyant cap over the sinking region of the conveyor, and throttling it. The ocean stopped transporting heat to the North Atlantic. Sea ice formed. The winds no longer carried heat to Europe. And so the Younger Dryas suddenly began, strewing delicate, little white flowers, and glacial cold, all over a continent that had so lately been shed of it.

The scenario had an almost visceral appeal to a man who had grown up in Oak Park, Illinois, and had driven his Soap Box car many times, albeit unknowingly, over the almost imperceptible continental divide that today separates the Gulf of Mexico and Atlantic watersheds. And there was another beautiful piece of evidence for it, almost a smoking gun. It came from the foraminifera contained in a couple of sediment cores from the Gulf of Mexico. Meltwater, like the ice it came from, is relatively poor in heavy oxygen-18. As long as most of the Laurentide meltwater was funneling to the Gulf, the forams in the bottom mud were bathed in "light" water and so were light too. In a certain layer of the sediments, however, the forams got abruptly heavier. Radiocarbon put a date on that change: it happened at the time the Younger Dryas was beginning in Europe. That was just when Broecker's scenario called for the meltwater to stop flowing into the Gulf, because it had been diverted into the Atlantic.

There remained the problem of explaining why the Younger

Dryas had ended, more than a millennium later, as suddenly as it had begun. Computer models helped Broecker with that. They not only confirmed that turning the conveyor off would dramatically cool Europe; they also suggested that if the conveyor's preferred state was "on," then it would naturally tend to switch on abruptly once a certain threshold of salinity had been crossed again in the North Atlantic. It seemed Broecker had come up with a plausible story, a mechanism in the ocean that could account for the "mode shifts" in climate that Dansgaard and Oeschger had documented in the Greenland ice cores. And it was a simple story, just the kind Broecker likes.

It soon got more complicated, that being the sad way of the world in general and of climate in particular. One problem Broecker didn't have to worry about for long was that a conveyor belt shutdown did not readily explain Oeschger's rapid CO_2 fluctuations—the observation that had prompted Broecker to think about the conveyor belt in the first place. Those fluctuations soon turned out not to represent real changes in atmospheric CO_2. After they failed to turn up in an Antarctic ice core, the Bern group decided the ones in the Dye 3 core had been caused by a chemical reaction between calcium carbonate dust and sulfuric acid in the snow. That released CO_2 and produced high concentrations of it in the ice.

Dansgaard's oxygen isotopes, however, did reveal real climate fluctuations—that much has been verified again and again in Greenland and elsewhere. The ice cores show, in fact, at least a dozen other abrupt changes in climate before the Younger Dryas. They happened between twenty-five thousand and fifty thousand years ago—during the Ice Age but before the Last Glacial Maximum. These "Dansgaard-Oeschger events" were similar to the Younger Dryas in duration and abruptness, but in one way they were quite different: they were abrupt warming rather than abrupt cooling events. The climate in Greenland climbed halfway out of glacial cold, stayed that way for seven hundred years or so, then plunged back. It was hard to see how the draining of Lake Agassiz, even if it were somehow to have happened a dozen times, could explain both warmings and coolings.

Nor were D-O events the only kind of weirdness the last ice age had to offer; they were just the spikes that stood out in the ice cores. Deep-sea cores from the northern Atlantic showed a different kind. In the sections of them dating to between 65,000 and 14,500 years ago, six distinct layers all had the great peculiarity of being almost entirely free of foraminifera shells, which are normally ubiquitous. What they contained instead were grains of continental rock that were too large to have been carried out into the middle of the ocean by winds or currents—they had to have been rafted out by icebergs. Some of the grains were around 2.7 billion years old, which is the age of the Canadian continental shield. Four of the six layers contained bits of limestone that another colleague of Broecker's, a geologist named Gerard Bond, traced to the mouth of Hudson Bay.

It was well-known that the bottoms of icebergs can be dirty with bits of rock that the mother glacier ground off the land, and that the icebergs drop this debris at sea as they drift along and melt. But these layers of debris stretched across the entire North Atlantic. They were as much as a foot thick near Hudson Bay and tapered to an inch or less near England and France; each layer contained as much as a trillion tons of sediment. Apparently the layers had been deposited by armadas of icebergs that had been launched in a relatively short time. These Heinrich events, as they are called after the German researcher who first identified the debris layers, did not coincide with the warm D-O events—they happened in between, when the climate was especially cold. It seems that every now and then a whole section of the Laurentide ice sheet, a great dome of ice centered on Hudson Bay, simply collapsed catastrophically.

Finally, as more people started to look into the matter, it became clear that the Younger Dryas was not just a European cold snap—and that the climatic effects of D-O and Heinrich events also were more widespread. When he first started working on the subject, Broecker had a young colleague, a woman named Dorothy Peteet, with whom he occasionally collaborated—for instance, they had worked well together one day in the Lamont parking lot, when they jacked up the back of their friend George Kukla's car and placed it on cinder blocks so that the wheels spun when he tried to drive off. This enraged the normally unflappable Czech, who was entertaining a distinguished Chinese loess expert. (Did people in China play practical jokes? Broecker asked, trying to lighten up an awkward moment. "Only small children," replied the visitor.)

Peteet had acquainted Broecker with the European pollen record of the Younger Dryas, but then she had the idea of looking for it in a swamp just across Route 9W from the Lamont lab. She found it there too. Later, after the Younger Dryas had become a cottage industry and Peteet herself had edited two learned volumes on the subject, to the point of being sick of it, she tried to escape by switching her research site to Kodiak Island, Alaska. She promptly found it there as well—the familiar brown-gray-brown layering of sediment. And with the right dates too.

Either the Younger Dryas or D-O events or Heinrich events, and sometimes all three, have since been identified in climate records from many parts of the world. They have been seen in sediments from the Santa Barbara Basin, from the Cariaco Basin off Venezuela, and from the Arabian Sea. They have been seen in the stalagmites of a cave in China. They have been seen in an ice core extracted from a glacier on a high peak in the Peruvian Andes. The Greenland ice itself contains evidence of the global reach of these abrupt changes, particularly the D-O events; the ice laid down in their warm phases contains lots of methane, which suggests that tropical wetlands, one of the atmosphere's largest sources of methane, were expanding at the same time Greenland was warming. During cold intervals on Greenland, on the other hand, the ice shows spikes in the amount of dust blown off the deserts of China, half a world away.

How all this fits together is unclear, but something more than just a shutting down of the conveyor belt is going on. Broecker has for two decades now been trying to fit together the pieces of the puzzle. During that time "abrupt climate change" has become a blossoming academic subspecialty. Hundreds of scientists regularly get together for days at a time to talk of nothing but. That talk can get incredibly arcane. Even an initiated listener may sometimes have trouble understanding why it is so important to know whether this particular wiggle in that particular isotope curve from a Chinese cave coincides with that other wiggle in the same curve from Greenland. But it does matter. Tracking the sequence of past climate shifts in time and their pattern in space is the best approach to figuring out the mechanism underlying them. And understanding the underlying mechanism of past abrupt changes is the only way to predict when they are likely to happen again. That realization never leaves the room for long, even when the discussion plunges into the most recondite mysteries of an isotope ratio.

EIGHT

Conveyor Jams, Climate Lurches

One thing that has become clear is just how amazingly abrupt some of the past changes have been. The cores drilled through the Greenland ice sheet after Dye 3 proved that. Dye 3 was not the best place to drill, and Dansgaard and his colleagues had known that; it was just the cheapest place, because there was already a military base there. But it was so far south that the snow sometimes melted in the summer, which blurred some of the annual layers in the ice and also helped generate Oeschger's spurious CO_2. Furthermore, a radar survey had revealed that the terrain underneath the ice at Dye 3 was extremely rugged, which meant one couldn't be sure that the bottom layers of ice—the ones that dated from the Ice Age—had not been folded or distorted. In the late 1980s, as he was still fleshing out the conveyor belt theory, Broecker helped browbeat the National Science Foundation into paying for another ice core; meanwhile Dansgaard and Oeschger organized funding on the European side. The result was that between 1989 and 1991, European and American research groups drilled two separate cores, around seventeen miles apart, through the very summit of the ice cap in central Greenland. Snow never melts there, and the underlying terrain is a plateau. Those two cores, the American one, called GISP2, and the European one, called GRIP, transformed climate science.

One agent of that transformation has been Richard Alley, now a professor at Penn State, who as a newly minted PhD was on the American team at GISP2. You cannot penetrate far these days into the world of climate research without encountering Alley's distinctive and high-energy presence. He is on the short side, and his gait is stooped and hurried; a bushy red beard and big glasses frame mobile eyebrows and eyes that often look just past you, as if his brain were unconsciously avoiding the drag of anchoring itself too tightly to a slower one. What makes him so popular though, among his students as well as his colleagues, who are used to hearing him give the inspiring opening talk at a conference, is precisely the generosity, as well as the skill and the slightly whacked enthusiasm, with which he cuts through the complexities. In a lovely book called *The Two-Mile Time Machine*, Alley tells the story of GISP2, starting with the first summer he spent on the ice sheet, prospecting for a drill site and living in a tent with two other guys. Vivid as the book is, it omits a few details—such as that Alley spent that first trip walking around with large slabs of foam rubber strapped to his feet, because he had been misinformed about just how cold it could be. Or that there was no outhouse.

That last problem was rectified by the time drilling started at GISP2 in 1989; *The Two-Mile Time Machine* includes a photo of the project's john, filled with a snowdrift after its door blew open. Even more alive in Alley's memory is that summer day in 1991 when the core section containing the end of the Younger Dryas finally made its way into the science trench. After it had been sawed in half lengthwise, it went first to a researcher named Kendrick Taylor, who dragged a pair of electrodes along the flat surface; the current between the electrodes rose and fell, tracing a spiky line on an oscilloscope, depending on whether there was a little or a lot of current-impeding dust in the ice. That was an approximate indicator of the past temperature—dusty periods were cold and dry.

The Younger Dryas was the most recent period of deep, prolonged cold. As they worked their way down the core, through the long, placid sameness of the Holocene, the GISP2 team knew it was coming. They had even jumped over some boring Holocene to get to it; Broecker's theory had endowed the Younger Dryas with special significance. But as they worked their way backward in time, they did not know just how fast the cold snap would be upon them.

"We're darn near a mile down at that point," Alley says. "And

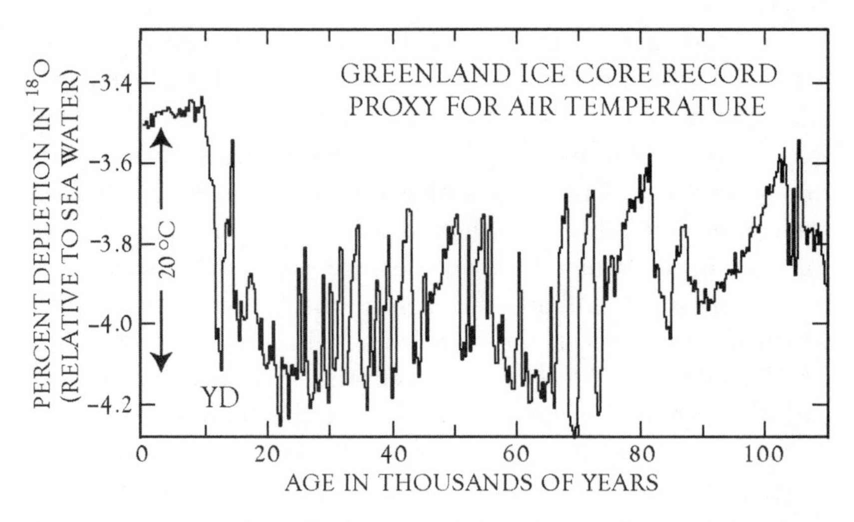

Oxygen isotopes in a Greenland ice core, GISP2, show evidence of large, abrupt swings in air temperature throughout the last ice age, notably at the beginning and end of the Younger Dryas period (YD).

the signal fell off a cliff, basically. He's getting into older and older ice, and it's saying, 'Just like everything else. Just like everything else. Just like everything else. Whoa—the world changed!' And then it just stayed down."

The isotope measurements from Camp Century and Dye 3 had shown that the warming at the end of the Younger Dryas had been abrupt. But at the time resolution of those measurements was not very fine; *abrupt* meant "less than a century." The GISP2 core, on the other hand, had beautiful, distinct annual layers. Alley was the team's expert at recognizing them; while still an undergraduate at Ohio State he had published a paper on how snow is transformed into ice in an ice sheet. In the science trench at GISP2, as the core sections went by on his light table, he counted the bands of summer bubbles and recorded them in his notebook, one by one—thousands of them, all the way back to the end of the Younger Dryas, 11,500 years ago. That's how the team knew they were at the Younger Dryas. After Taylor saw the dramatic dust spike, and everyone had crowded around him, Alley put that core section on the light table. At the same point where Taylor had seen his spike, Alley saw that the ice layers were suddenly half as thick as the ones he'd been looking at in the higher, Holocene core sections. Looking deeper down the core and into time, he saw that the arid period of low snowfall had persisted for the entire Younger Dryas. At the end of that period, it had suddenly started snowing twice as much in Greenland, and the air had gotten several times less dusty. Most of that change had taken place in just a few years.

The beginning of the Younger Dryas had been slightly more gradual, but only slightly—a matter of a few decades at most. At that point, 12,900 years ago, Greenland had suddenly gotten colder and drier and stayed that way for more than thirteen centuries; then it had suddenly gotten warmer and wetter, and it has stayed that way ever since, relatively speaking. (Compared with most of the world, Greenland is not what you would call warm or wet.) Just how big were these changes in climate? GISP2 shed new light on that question too. Though Dansgaard had referred to his oxygen-isotope oscillations as "violent" and "drastic," he hadn't attached particular temperatures to them, because he had no way of knowing whether the calibration he had established in the modern world, between air temperature on the one hand and the isotope ratio in rain and snow on the other, had held in the same way during the Ice Age.

Researchers at GISP2 found two ways around that problem. The one that Alley and his then student Kurt Cuffey developed was the more intuitive: they stuck a thermometer down the borehole. It may seem surprising, but the ice on Greenland does not have the same temperature through its entire two-mile thickness. It has thermal inertia; it has memory. It remembers the temperature that prevailed when it formed, thousands or tens of thousands of years ago, just as a roast you pull from the freezer and then stick in the oven (to use an example of Alley's) preserves at least for a while, in its icy interior, a memory of its past frozen state. The ice a mile down is much colder than the ice on the surface, because it was laid down twenty thousand years ago at the Last Glacial Maximum. While the air over Greenland has warmed a lot since then, the warmth has not yet fully penetrated a mile down. A temperature reading from that depth does not directly tell you the LGM air temperature—things are more complicated than that—but it is an indicator of the temperature, just as the oxygen isotopes in the ice are an indicator. By combining the two measurements, you can determine the past air temperature—that's what Cuffey figured out how to do.

Meanwhile a former student of Broecker's, Jeff Severinghaus of the Scripps Institution of Oceanography, was developing a far more subtle way of measuring the amplitude of past warming events such as the end of the Younger Dryas, or the one immediately after the LGM. Severinghaus's method is far too subtle and intricate, in fact, for us to explain here. Suffice it to say that it involves measuring isotopes of nitrogen and argon in the trapped air bubbles—and that it yielded almost precisely the same results as the borehole measurements. That suggests you can believe both. At the Last Glacial Maximum, the temperature on Greenland was between twenty-one and twenty-three degrees Celsius, or thirty-eight to forty-one degrees Fahrenheit, colder than it is today. During the Younger Dryas it was sixteen degrees Celsius, or twenty-nine degrees Fahrenheit, colder.

Large as it is, that temperature drop makes the Younger Dryas seem less ferocious than it really was. The temperature is an average over a whole year—isotope and borehole measurements blur the seasonal differences. A few years ago, though, Alley and George Denton found strong evidence that winters in the Younger Dryas were exceptionally harsh. In the summer of 2003, the two researchers sailed into Scoresby Sound with Gary Comer on his yacht *Turmoil.* (Two years after his cruise through the Northwest Passage, Comer was actively supporting climate research, to the extent of using his own boat for it.) Scoresby Sound is on the east coast of Greenland, around three hundred miles due east of the GISP2 and GRIP drill sites. It is an austerely gorgeous bay, nearly devoid of life but traversed by dense flocks of icebergs, which feed into it from long fjords. The fjords reach nearly two hundred miles inland; one of them, Nordvestfjord, is nearly a mile deep. They were carved during the last ice age by glaciers that have since retreated to the heads of the fjords. At the Last Glacial Maximum, ice filled Scoresby Sound all the way out to the Greenland Sea.

What Denton and Alley found, however, after logging six thousand miles in Comer's seaplane and tramping around onshore a good bit as well, was that the subsequent resurgence of the glaciers during the Younger Dryas had not been nearly as strong as one might expect for an annual average temperature drop of twenty-seven degrees. The Younger Dryas moraines on Milne Land, an enormous island in the sound, extended less than eight miles past the farthest position occupied by the glacier in recent centuries. From those moraines and other features, such as glacial trimlines, the researchers could reconstruct, just as Denton and his team have done in the New Zealand Alps, the past position of the equilibrium snow line—the altitude at which a glacier loses as much snow as it accumulates. As a paleothermometer, the snow line measures the summer temperature, because the biggest influence on whether a glacier expands or retreats is how much of the previous winter's snow is lost to summer melting. In Scoresby Sound during the Younger Dryas, the snow line altitude dropped only two thousand feet or so; it would have fallen three times that much if the temperature had been twenty-seven degrees lower than today all year around. The reason it didn't fall that much, Denton and Alley figure, is that the summer temperature in Greenland only fell by around nine degrees. To drop the annual average by twenty-seven degrees, the winter temperature must have dropped by a whopping fifty degrees.

A similarly strong seasonal pattern has been documented from records in northern Europe; there too the Younger Dryas was mostly a winter event. Winter temperatures dropped an average of forty degrees Fahrenheit (twenty-two degrees Celsius) in Europe, based on the available evidence, whereas summer temperatures fell no more than around ten (six degrees Celsius). In Britain the difference between summer and winter temperatures was around fifty-five degrees Fahrenheit, thirty-five more than it is today. Basically, a green and pleasant maritime climate became, all at once, harshly continental. The most likely explanation is that the west winds that now bring the warmth of the Atlantic to Britain were no longer blowing over a warm ocean. Instead they were blowing over a vast frozen sea.

A frozen North Atlantic would explain the extreme winters in Greenland too. During the Younger Dryas, Arctic sea ice probably spread south past Iceland and Greenland, until its edge was a line stretching from Newfoundland to Normandy. Arctic foraminifera lived in those latitudes then, as Ruddiman and McIntyre documented decades ago, because the northern Atlantic then was like the Arctic now. How do you make the entire northern Atlantic so cold that it freezes over? Only one answer seems reasonable: you shut down the conveyor belt and stop the northward flow of heat from the tropics.

Paradoxical as it may sound, an earlier shutdown of the conveyor, and the spread of sea ice over the northern Atlantic, may have helped bring about the end of the last ice age—the abrupt "termination" that Broecker has been trying to explain since he first encountered it at Pyramid Lake fifty years ago. At the time he thought of it as "abrupt," anyway, but the meaning of that word in climate circles has changed as a result of the Greenland ice cores. Compared with the ninety thousand years it took the ice sheets to build up to the Last Glacial Maximum, the six millennia or so it took to get from there to the end of the Younger Dryas, and the relatively ice-free conditions of today, the termination does indeed seem abrupt. Compared with the spikes in the ice-core record, it was anything but. The termination ended with the Younger Dryas, included a couple of lesser spikes before that, and began, 17,500 years ago, with a massive Heinrich iceberg armada. Its timing and magnitude were determined by orbital changes in sunlight, by the ice sheets themselves, by the conveyor belt, and by atmospheric carbon dioxide. In 2006, Denton, Broecker, and Alley sketched a scenario of how all these actors might have worked together.

The changes in the first three millennia of the termination have been so perplexing that Denton and Broecker had taken to calling that period the Mystery Interval. The biggest mystery has been this: the oxygen isotopes in Greenland don't show any special change at the onset of the termination 17,500 years ago, and certainly not a warming. Indeed, they indicate that during the Mystery Interval the North Atlantic region actually got colder—on an annually averaged basis, that is. Yet 17,500 years ago is when mountain glaciers and ice sheets all over the world began their retreat from the Last Glacial Maximum. Denton has spent his career mapping that retreat. In New Zealand it was especially rapid, with the glaciers taking only a millennium to withdraw high into the Southern Alps. In the European Alps the retreat may have lasted the full three millennia of the Mystery Interval, but it began at the same time. The Laurentide ice sheet in North America also started to recede then.

The solution to the Mystery may be the same as the one for the Younger Dryas: a harsher contrast between the seasons. The North Atlantic region was indeed warming in summer, which is what you need to melt ice. The Milanković orbital curves indicate that the amount of sunlight striking the northern hemisphere during summer was climbing toward a maximum as the termination began. But it remained brutally cold in winter, because the northern Atlantic itself was frozen—it remained covered with ice at the same time as the ice on the adjacent land was melting. The cause of that was a shutdown of the conveyor, just as in the Younger Dryas. But the cause of the shutdown was different: instead of the sudden draining of a giant glacial lake, it was the launching of a vast iceberg armada from Hudson Bay. The Heinrich event occurred at the same time as the onset of the termination. All those melting icebergs deposited an enormous amount of freshwater in the northern Atlantic, which would have been ample to jam the sinking leg of the conveyor. Evidence from sediment cores confirms that the shutdown was close to complete.

The heat that the conveyor was no longer carrying into the North Atlantic did not vanish. It accumulated in the South Atlantic—and ice cores from Antarctica indicate that Antarctica, unlike Greenland, did start to warm 17,500 years ago. The warmer water in the South Atlantic, besides warming the air, would have melted the greatly expanded fringe of sea ice that surrounded Antarctica at the Last Glacial Maximum. And that would in turn have allowed deep waters to well up to the surface there, as they do today. Those deep waters are rich in carbon dioxide. A few years ago Ralph Keeling and his colleague Britton Stephens proposed that the expanded Antarctic sea-ice cap during the last ice age had been like a bottle cap that trapped CO_2 in the deep ocean, thus explaining why there was less CO_2 in the atmosphere then. Once the cap was off, CO_2 began to bubble out of the sea and the atmospheric concentration began to rise. The Antarctic ice cores show that rise began, like the worldwide retreat of the glaciers, around 17,500 years ago.

Once CO_2 began to rise, the whole planet began to warm. At some point the conveyor switched back on; the Greenland cores record a sudden warming 14,500 years ago. That ended the Mystery Interval and started a period known as the Bölling/Alleröd. Sea ice retreated from the northern Atlantic back into the Arctic, amplifying the warming—until Lake Agassiz found its new outlet to the sea, which shut down the conveyor again. During the ensuing Younger Dryas, sea ice once again spread south. Its effects were felt around the world, and not just around the northern Atlantic. A computer modeling study done by John Chiang of the University of California has shown that extensive sea ice in the Atlantic, even if it is there only during the winter, pushes the tropical rain belt well to the south, taking rain from some latitudes and giving it to others. That explains why the Younger Dryas shows up in sediments off Venezuela and in stalagmites in Brazil.

Its presence in caves in China is another matter—that probably indicates a failure of the monsoon. Sea ice in the northern Atlantic would have promoted longer, colder winters throughout Eurasia, including the Tibetan Plateau. Hot air rising over the Tibetan Plateau in summer is the engine of the monsoon, the way cold water sinking in the Atlantic is the engine of the conveyor belt. By stalling summer in Tibet, the Atlantic sea ice may have stalled the monsoon, with effects felt throughout the region. Everything is connected in the global climate system.

In this particular scenario of how things are connected, the shutdown of the conveyor was crucial: it froze the northern Atlantic, which had ripple effects around the world, and it melted the South Atlantic, which had far more than ripple effects—by pumping CO_2 into the atmosphere, it brought the planet out of the Ice Age. And yet there was nothing mechanical or predetermined about the termination. On the contrary it was somewhat random, especially in the way it began. The Ice Age had seen five previous Heinrich events, and none of them had killed the ice sheets. There had been several peaks in the Milanković sunlight curve and none of them had done the job. Maybe the Ice Age could end only when it was ready—when the Laurentide ice sheet had gotten so big, at the Last Glacial Maximum, as to be unstable, such that some nudge to the dome of ice that covered Hudson Bay caused it to collapse catastrophically and spatter the North Atlantic with icebergs. That's not a random event exactly, but it's not the sort of thing that is easy to foresee. If Broecker had been sitting in his office at Lamont not just 50 years ago, but 17,500 years ago, and if he had somehow been able to function as a climate theorist while buried under a mile of ice, he would not likely have predicted that the ice was about to go away.

Broecker has worn down many a No. 2 pencil working out his theories of climate change. He writes all his books and scientific papers in longhand. There is no computer in his office. Accepting that you're a dinosaur is the first step toward overcoming the condition; if you're also a scientist, the key step after that is keeping your brain open to other brains. Other researchers, many researchers, have built and run the computer models that show that Broecker's scenario for the Younger Dryas, say, could well be true—that a jolt of freshwater to the North Atlantic from Lake Agassiz could have stopped the conveyor belt, which would have caused the climate effects observed in sediment cores and ice cores and stalagmites and moraines. Other researchers, many more still, have retrieved and analyzed those samples of the real world. Broecker's favorite way of staying in touch with these people, some of whom are at Lamont, is just to leave the door to his office always open. It's next to the front door and to the mailboxes of the geochemistry building, and so people are always passing by and looking in. His least favorite method is e-mail;

Moanna St. Clair, his secretary, prints it out for him, and Broecker scribbles terse replies on the printouts. On the phone, which he much prefers, he is far more discursive, though his trademark is still to hang up without saying goodbye, as if life were a single, never-ending conversation. Whereas really it is just science that is something like that—a conversation among the people who do it, a never-ending conversation between ideas and the real world they represent so imperfectly.

In March 2005, one of the people Broecker was keen to get a call from was Tom Lowell, a glacial geologist at the University of Cincinnati. Lowell is a longtime collaborator of George Denton's and of similar shape, which is to say big. Early that month, Lowell was in a helicopter hopscotching around the boreal forest of northern Alberta. A laptop perched on his knees displayed a photo taken by the space shuttle, which he was using to guide the pilot around the uncharted wilderness. In the slanting light of morning, the endless forest of stunted trees—trunks not much thicker than a man's arm, cones of foliage not much wider than the trunks—looked almost like undulating fields of grain, and just as featureless. But when you looked closer, the details emerged. Stands of bare, silver-gray poplar looked like patches of mist; here and there a dark pine or spruce rose out of them like a fairy-tale castle. Cloud shadows added to the moodiness of the landscape—except, of course, when the whole thing was obscured by driving snow, as it sometimes was. On those days the frozen lakes and ponds that Lowell was hunting for were harder to find.

Yet he and his colleague Tim Fisher of the University of Toledo were managing to set down on half a dozen a day. They would unstrap the drill equipment from the sides of the helicopter, hump it through knee-deep snow to some plausible spot, cut a hole through the ice with a hand auger as if they were going fishing, and spend half an hour or so setting up the drill while the helicopter took off on other business. Then, after perhaps a moment's pause to appreciate the gaping silence of the place, they would drive the coring device noisily through the ice and water and into the mud on the bottom. The cores they were using to test Broecker's theory of the Younger Dryas were shaped like Broecker's pencils, but they were heavier. "We are the worker bees," said Lowell one evening, while soaking away the day's ache at the Keg Steakhouse in Fort McMurray. The description more or less fits Fisher, a man of average build, thoughtful speech, dry wit, and in wintertime, reddish beard. Lowell himself seems more of a worker bear. In college at Rensselaer Polytechnic Institute he set records in the hammer throw.

Fort McMurray is a boomtown: it sits on the edge of the Athabasca tar sands—vast deposits of low-grade oil that, as the price of a barrel has soared, have attracted huge investments from the major oil companies. What Lowell and Fisher were drilling for, though, was evidence of a cataclysmic flood. This flood, at the close of the last ice age, stripped tens of millions of years' worth of sedimentary rock off the tar sands, leaving them so close to the surface that the oil companies can mine them with steam shovels in open pits. For his PhD in the early 1990s, Fisher had charted the path of the flood. Southeast of Fort McMurray, he showed, the flood had cut a valley three-quarters of a mile wide for what is now a piddling river, the Clearwater. And in the mine pits north of Fort McMurray, where the Clearwater joins the Athabasca River, Fisher found sheets of gravel that had clearly been deposited by a flood—they contained boulders four or five feet across.

Was Fisher's flood the same as Broecker's flood, the one that supposedly drained Lake Agassiz and triggered the Younger Dryas? That was the question Lowell and Fisher had come to Fort McMurray to answer. Broecker's original idea had been that the Agassiz flood had gone east through Lake Superior and down the St. Lawrence. But Lowell and Fisher had already been to the northern shores of Lake Superior. From the moraines there they had concluded that the flood couldn't have gone thataway; the whole area was still covered by the Laurentide ice sheet during the Younger Dryas. Perhaps instead it had gone to the northwest, along the edge of the retreating ice sheet—through the Clearwater, Athabasca, and Mackenzie river valleys to the Arctic, and then on to the Atlantic. If so, then the key floodgate, judging from the topography, was the Athabasca valley north of Fort McMurray.

From their base camp at the Radisson in Fort McMurray, Lowell and Fisher and their two graduate students were fanning out across that area, trying to date the retreat of the ice sheet—to find out whether it had opened the floodgate at just the right time, around thirteen thousand years ago, to cause the Younger Dryas. As the ice sheet retreated to the northeast, it left behind a series of moraines, and behind the moraines it left depressions it had gouged out of the land. Those depressions soon filled with water to become lakes. By drilling into the lake sediment and dating the oldest layer—lakes invariably contain life, which means the sediment contains organic material suitable for radiocarbon dating—Lowell and Fisher could determine when the lake had formed. Presumably that was not long after the ice sheet had left, thus perhaps opening an outlet for Lake Agassiz.

That March they cored twenty-seven lakes in all. Once the dates came back from the radiocarbon laboratory, a consistent pattern emerged: all the lakes around Fort McMurray apparently formed at least a millennium after the Younger Dryas began. The ice sheet had retreated and a flood had indeed passed that way—but according to Lowell and Fisher it was at least a thousand years too late for Broecker's theory. "We're not finding evidence to support the theory, and all the evidence we find is pointing the other way," Lowell said at the end of the field trip in March 2005. When Broecker got that news, he was certainly not pleased. It was embarrassing. It spoiled a good story—a story that his scientific peers closely identified with him. But that's the way things go in science. The important thing is not to identify yourself too closely with your own theories—not to get too attached.

It's also important, however, not to throw your baby out with the floodwaters. There is no question that Lake Agassiz existed during the Ice Age and that it is not there now. There is solid evidence from sediment cores that it once drained into the Gulf of Mexico but that it stopped doing so around the time of the Younger Dryas. And there is solid evidence from ancient shorelines that the water level in the lake dropped by 130 feet sometime around the Younger Dryas. (The precise timing is uncertain.) All that water had to go somewhere; we just don't know where at the moment, or exactly when. The best guess remains that it made it to the Atlantic somehow—maybe it tunneled under the ice sheet without leaving a trace—and that it did indeed cause the Younger Dryas. That remains the best guess because no one has come up with a better one. And one final piece of evidence supports the Lake Agassiz scenario: judging from the long record in Antarctic ice cores, no abrupt cooling like the Younger Dryas seems to have interrupted the warming at the end of previous ice ages. That suggests the Younger Dryas was a random, freak event—which fits a scenario that depends on a coincidence between the shape of a receding ice sheet and the topography of the land underneath to trigger it.

There is a hierarchy of certainty and of importance to our ideas about abrupt climate change: we are more certain about the more important ones. We are not certain that a Lake Agassiz flood triggered the shutdown of the conveyor that caused the Younger Dryas; maybe it was another iceberg armada that provided the freshwater jolt, or maybe the conveyor was stopped by something other than freshwater. We are far more certain that a conveyor belt shutdown did indeed cause the Younger Dryas and other abrupt climate changes during the Ice Age; sediment cores from the deep Atlantic document the shutdown, and computer models confirm its effects on climate. Finally, we are close to dead certain, from the Greenland ice cores and other climate records, that abrupt climate changes did actually occur during the Ice Age, whether or not they were caused by the conveyor belt. The existence of abrupt climate change is by now more of an observational fact than a theory. Along with the fact that CO_2 is accumulating in the atmosphere and inevitably warming the planet, it is the most important thing we have learned about climate in the past half century.

For most of that half century Broecker has been sitting in the same office at Lamont, the one he inherited, along with its furniture, from his PhD adviser, Larry Kulp. But he will soon be moving: a new geochemistry building, paid for by Gary Comer, is rising outside his window. There will be many mementos to pack—the photographs of the dozens of graduate students he himself has had over the decades, along with their own PhD theses; the life preserver stolen from the *Knorr* during GEOSECS; the mannequin of a bearded lady—or in some cases perhaps not to pack.

One artifact that is sure to make the journey is the mangy

stuffed snake that Dorothy Peteet found by the side of the road years ago and deposited in Broecker's office. He liked it so much he hung it in front of the windows. Around twenty feet long and fuzzy, it is aqua blue with pink spots. Its head, with prominent black nostrils and a flickering red tongue, nearly touches Broecker's desk. The sign around its neck reads, "I am the Climate Beast and I am Hungry."

NINE

Why Worry?

Lifting off from the little airstrip, the helicopter banks north, up the Tasman Valley. Below, the silver braids of the river wind through glacial mud, punching brown clouds of it into the powder blue of Lake Pukaki. Ahead lies the Tasman Glacier, source of the river; beyond lie the ice-covered peaks of the Southern Alps, source of the glacier. It is the longest glacier in New Zealand, but it used to be far longer and thicker still.

George Denton's voice comes over the headphones. The helicopter is climbing along the eastern flank of the valley, which is green but unwooded. During the ice ages, as the glacier repeatedly filled the valley and then drained from it, it left behind shelflike and easily visible moraines of rubble on the steep slopes, the way bathwater draining from a tub leaves rings of dirt. The moraine nearest the top, says Denton, is 140,000 years old or so; it dates not from the last ice age but from the one before that. Farther down the slope, but still more than half a mile above the valley floor, is the moraine from the Last Glacial Maximum, which Denton and Bjorn Andersen have traced thirty miles south of here to the southern tip of Pukaki. There it dams the lake that now sits where the glacier's snout sat at the LGM. After the ice retreated back up the valley from that position, it underwent one more great surge, over several millennia, that ended with the Younger Dryas. A moraine near the bottom of the slope records that Late Glacial Advance.

The next day, Denton leads his companions on foot into the Tasman Valley to get a closer look at yet another moraine—a steep one that crosses the valley floor. The Tasman River spills from a gap in its center, heading toward Pukaki; behind the moraine lies Tasman Lake, which sits on the snout of the present glacier, on a bed of ice that is still more than six hundred feet thick. Unlike the ice-age moraines high up the valley walls, this one is gray and mostly unvegetated, because plants have not had much time to colonize the rubble and turn it into soil. In 1862, when a European explorer first sketched the Tasman Valley, the glacier occupied this same position. But from the top of the moraine then, you would have been looking up at a wall of ice; now you look down, far down a gray slope to a lake of meltwater. Scattered gray icebergs dot the lake. A yellow powerboat laden with tourists zips about. The barren gray slope, several hundred feet high, surrounds the lake on all sides and continues along the walls of the valley, immediately above the glacier—a vivid marker of vanished ice.

The nineteenth century in the northern hemisphere was the end of a period called the Little Ice Age, which had begun in the fourteenth century or so. Compared with the real ice age, the Little Ice Age was a mere cold snap, but glaciers recorded their greatest advance since people had begun to record such things. Glacial history is not as well known in New Zealand, but in the nineteenth century the glaciers there, too, were much more extensive. Since then the volume of ice in the Southern Alps has diminished by nearly half. The Tasman glacier is unusual in that it still touches the moraine it built in the nineteenth century; it is insulated from the summer sun by a thick blanket of rubble that has tumbled onto it from the surrounding mountains—its snout looks more like a moonscape than a field of ice. But as the bare moraine slope around it shows, the glacier has dramatically diminished in thickness, by hundreds of feet. And since the 1960s it has done so at an accelerating pace.

Weather records indicate that the temperature in New Zealand has risen by about a degree Celsius in the past century, higher than the global average rise of six-tenths of a degree. Most of the rise has happened since the 1950s. Only later did lakes begin to appear at the snouts of the largest glaciers in the Southern Alps, the ones like the Tasman that plunge to low altitudes and fill valleys. Tasman Lake did not exist before 1965. Once such lakes are born, however, they tend to grow rapidly, eating away at the glacier, breaking icebergs off it—a glacier loses ice much faster that way, by calving, than by simply melting at its surface in the summer sun. Between 1986 and 2001, Tasman Lake tripled in size, to well over a square mile. It hasn't stopped growing since.

The glaciers of New Zealand are dramatic evidence of man-made global warming—and also of the complexity that has made it so easy for people to be skeptical of the reality of the problem. Between the 1970s and the mid-1990s, as Tasman Lake was growing at the expense of its receding glacier, other glaciers in the Southern Alps were actually advancing again. The size of a glacier depends not only on how much ice melts during summer at low altitudes, but also on how much snow accumulates during winter at high altitudes. In the late 1970s, the prevailing winds over the South Island of New Zealand shifted from the northeast to the southwest—a cyclic change related to sea surface temperatures in the Pacific—and those wetter southwest winds dumped more snow onto the Southern Alps, causing some glaciers to surge. But in recent years, as the climate has continued to warm, all the New Zealand glaciers have resumed their long-term retreat. In the coming century, as the warming moves the snow line relentlessly up the mountainsides, the glaciers are doomed to retreat with it, and the meltwater lake will grow.

Yet as skeptics often point out, the retreat began more than a century ago. It began almost exactly at the time when Svante Arrhenius was discovering the possibility of a man-made greenhouse effect—but before we had created much of one, because we had still added only a small amount of carbon dioxide to the atmosphere. Thus the glacial retreat began as a natural climate fluctuation. The effect of our emissions has been overlain on that fluctuation. Only in recent decades has it clearly taken over—has the "signal" of man-made climate change begun to emerge from the background "noise" of natural variability.

There can be no mathematical proof that the signal is now detectable, but there is proof that meets the legal standard of "beyond a reasonable doubt." No one today doubts that carbon dioxide added to the atmosphere warms the planet; it's a matter of established physics, like Newton's laws. The amount of carbon dioxide in the atmosphere is now higher than it has been at least in the last past seven hundred thousand years—the length of the longest Antarctic ice core—and probably in the last few million years or even tens of millions of years. The warming of the past three decades has coincided with the surge in our emissions that Dave Keeling documented. When computer models of climate include the effect of that carbon dioxide, they reproduce well the observed spatial pattern of the warming—that it is more pronounced at high latitudes than in the tropics, and that the warming in the lower atmosphere actually gives way to cooling in the stratosphere, which is no longer receiving the infrared heat radiation that the added CO_2 is trapping near the surface. But when those same models include only natural effects on climate, they can't reproduce the warming of the past three decades at all. Changes in the output of the sun have been minimal at best. Volcanic eruptions have been frequent and should, if anything, have been cooling the planet. Yet it has continued to warm.

Anyone who visits the Tasman Valley can see the Ice Age moraines high up the valley wall, vivid evidence of the extent of natural climate variability. But in the state of the glacier today he is also seeing the effects of man. "When skeptics ask me, 'Is global warming real?' I tell them about this," says Denton, sitting on the moraine above Tasman Lake. "I've been to glaciers all over the world. This is happening all over."

And so the debate on the reality of global warming is pretty much over now. The interesting questions are how bad a problem we have, and what to do about it.

There are many reasons to be worried about global warming; it is changing the whole world. In 2007, the Intergovernmental Panel on Climate Change released, in lengthy installments, its fourth report, in which six hundred scientists from forty countries reviewed the evidence that man-made global warming is already happening, the effects it is already having, and the outlook for the future. The third IPCC report, released in 2001, had already included a long list of worrisome impacts. The tone of the new report was more urgent.

Compiling this fourth IPCC report, researchers using different computer models to simulate future climate have reached greater agreement on just how sensitive Earth's climate is to carbon dioxide. This sensitivity is expressed as the amount of warming predicted for a doubling of atmospheric CO_2, from the preindustrial level of 280 parts per million to 560 parts per million. If developing countries develop as rapidly as they should to alleviate poverty—the worst-case or best-case scenario, depending on your point of view—and if 85 percent of the world's energy continues to come from fossil fuels, CO_2 could double as soon as the middle of this century. The best estimate now is that the global average temperature would rise by around three degrees Celsius, or five degrees Fahrenheit, as a result.

Plants and animals are responding to the warming already in a completely unsurprising way, just as they did at the end of the last ice age: they are shifting their range toward the poles and higher up the mountains, in an effort to remain in the temperature zone they are adapted to. The shift has been observed in a wide range of species. It has been especially well-documented in European butterflies; Camille Parmesan of the University of Texas and her colleagues found that of the thirty-five species of butterflies they looked at, twenty-two had shifted their range northward during the twentieth century, by anywhere from 35 to 240 kilometers. Only one species had shifted southward. Meanwhile in the Alps as in the Rockies, trees and grasses have been moving upslope, colonizing areas that once were above the tree line—which is not all bad, of course. But it can be very bad indeed for the plants and animals that had retreated to the cold and treeless high peaks after the Ice Age and are now being displaced.

Perhaps the most dramatic example of global warming's impact on animals has been the plight of the polar bears, which hunt seals from sea ice in the Arctic. Satellite observations have shown that sea-ice coverage has declined by 30 percent in the past quarter century; by midcentury the Arctic may be entirely ice-free in summer, a condition it has not seen for at least a million years. Polar bears on Hudson Bay and elsewhere in the Arctic are not eating as well as they used to, and there have even been reports that some of them are drowning as they try to cross large stretches of open water between ice floes. At the end of 2006 the Bush administration proposed adding polar bears to the U.S. government's list of threatened species, "threatened" being a step below "endangered." *The Washington Post* account of the decision said it "could have an enormous political and practical impact" on U.S. policy on global warming.

That is questionable. Televised pictures of clubbed and bloody seal pups, and highly publicized protests of the Canadian seal hunt, have never managed to end the hunt, even though it is an insignificant part of the global economy; in 2006, Canadian hunters killed more than three hundred thousand seals, about as many as when the protests started in the 1960s. So it seems unlikely that pictures of emaciated polar bears and even the threat of their extinction will convince the world to revolutionize the very foundation of the global economy—and it will take a revolutionary change to the energy supply system just to slow down global warming. (Even if we were to stop all emissions of carbon dioxide today, the planet would still warm by another 0.6 degree Celsius or so, as the ocean radiated to the atmosphere some of the excess heat it absorbs.) Human beings are not going to make that kind of effort, in our opinion, to save polar bears or any other wild organism. More powerful appeals to self-interest will be needed.

One might think that evidence that global warming actually kills people might do the trick. The World Health Organization has estimated that global warming is already, right now, causing as many as 150,000 excess deaths a year. The figure, first published in 2002, is highly uncertain and might sound alarmist—but the European heat wave of 2003 did lend it some credence. That summer in France and Italy was completely outside the envelope—it was by far the warmest since temperature records began in 1851, and probably the warmest since 1500. It caused somewhere between twenty-two thousand and forty-five thousand excess deaths. One can't say that global warming caused the heat wave, any more than one can conclusively attribute a particular lung cancer to the victim's smoking two packs of cigarettes a day. But one can say, as Peter Stott of the U.K. Met Office and his colleagues did after modeling the summer of 2003, that global warming had made such a heat wave at least twice as likely as before—and that if CO_2 emissions continue to grow on their present course, such heat waves will be more common than not by midcentury in Europe and will be considered freakishly cold by century's end. Similarly, there is evidence, albeit more controversial and debatable, that global warming is increasing the spread of ailments such as dengue fever, malaria, and diarrhea.

Yet talk of such effects has been going on for years without inspiring much action on the part of the governments that matter—that of the United States, for instance. One reason is obvious as soon as you look at the color-coded world map in the WHO study, showing the number of deaths attributed to global warming in the year 2000. Almost none of them occurred in the developed CO_2-spewing countries of the north; almost all occurred in poor tropical countries, above all in sub-Saharan Africa, because that's where the diseases promoted by warming occur. The other reason for inaction is that you can always argue that it is cheaper and more effective to attack health impacts directly than to attack global warming. If you want to prevent malaria deaths in Africa, you're more likely to succeed immediately by distributing sleeping nets impregnated with pesticide than by trying to wean the world from fossil fuels. If you want to prevent heat deaths in Europe, the first thing to do is what France in fact started to do in 2004—launch a crash program to make sure nursing home residents have access to air-conditioning. (Fortunately most French electricity comes from nuclear power, which doesn't emit CO_2.)

Much the same thing could be said about the effect of global warming on hurricanes, a topic that was hotly debated in the aftermath of the destruction of New Orleans by Hurricane Katrina. Whether or not the effect has already been observed, it stands to reason that global warming will intensify hurricanes one day, as it warms the tropical seas from which they draw their energy. As the storms grow more intense, the damage they cause when they make landfall will rise too. But if you want to reduce that damage, your first step should not be to try to remake the global energy system—it makes more sense first to remake the government policies that have allowed and even encouraged massive development of coastal zones, thus putting more people and their buildings in the path of hurricanes.

To our minds, the best arguments for doing something about global warming are the ones that concern impacts that are most clearly and directly tied to the warming, that have a huge economic cost, and that can't be prevented more easily in some other way. To those three criteria we would also add a fourth: if the impact has the potential of coming at us fast—of being an abrupt change like the ones that are so abundant in the paleoclimate record—then there is far more reason to be worried, because we might not have time to adapt to it.

Could, for instance, the conveyor-belt circulation suddenly stop, the way it apparently did at the start of the Younger Dryas and the Dansgaard-Oeschger events? And would that stoppage cast a frigid pall over Europe and eastern North America, even in a globally warming world? The scenario has gotten a lot of press, and Broecker obviously bears some responsibility for that. In 2004 the film *The Day After Tomorrow* pushed the scenario to its absurd if entertaining limit, burying much of the United States under an ice sheet within days of a conveyor shutdown, with the American population forced to flee south to Mexico to escape the advancing ice.

That had Broecker squirming in his seat at the multiplex. As evidence for the effects of a conveyor shutdown in the past has grown ever stronger, he himself has become less worried about such an event in today's world. In principle, global warming might indeed dump enough freshwater into the North Atlantic to jam the conveyor, either by melting ice in Greenland or by increasing rainfall and the flow of rivers that empty into the Arctic and northern Atlantic. There is even some evidence that the process is under way already; the northern Atlantic has indeed freshened in recent years, and there have been preliminary reports of an actual slowdown in the currents. But the best calculations still suggest that the conveyor is more likely to slow rather than come to a complete halt—and that by the time it does slow significantly, the planet will have warmed so much that whatever cooling the conveyor slowdown triggers in the countries around the North Atlantic will more likely be felt as a relief than as a catastrophe. The huge worldwide effects triggered by conveyor shutdowns during the Ice Age, it seems, could only happen when the North Atlantic was covered with sea ice. Global warming itself pretty much eliminates that possibility today.

From Al Gore on down, environmentalists keen to spur action on global warming sometimes argue that it is a threat to Western civilization. The motive is reasonable, but the argument strikes us as tenuous; Western civilization seems more resilient than that. But the evidence from the past suggests that global warming does indeed have the potential to do grave harm, in two ways in particular. Both have to do with the way the warming will redistribute water around the planet. First, it could lead to prolonged, catastrophic droughts in certain regions, such as Africa and the American Southwest—and the paleoclimatic record suggests such droughts might begin suddenly and with little warning. We'll talk about that risk in chapter 11. The second danger, virtually certain, is that global warming is going to melt a lot of ice and thus raise the level of the sea—slowly, we hope, but right now no one knows for sure.

TEN

Ice Melts, Sea Level Rises

The big melt has already begun. As Denton said, the New Zealand glaciers are not a special case. Mountain glaciers all over the world are retreating. Glacier National Park has lost nearly three-quarters of its ice-covered area since 1850, and its glaciers will probably vanish entirely before the middle of the century if temperature continues to increase at the present rate. The best documentation comes from the Alps of Europe, where glacial retreat has also been general and headlong, and where ski resorts at lower altitudes are beginning to panic about their future. Nineteenth-century photographs show the imposing mound of the Rhône Glacier way down in its valley, within a few hundred yards of the hotel built there to accommodate the first tourists; since then the glacier has retreated nearly a mile and a half, out of the lower valley, up a steep sixteen-hundred-foot slope, and beyond the lip of a rock ledge. It is now barely visible from below. Up on the ledge, the owners of the Hotel Belvedere have lately started covering the edge of the glacier in summer with tarpaulins to try to slow the shrinkage and to protect the tunnel they dig each summer for tourists.

The ice retreat has been especially dramatic at mountain glaciers in the tropics. A glaciologist named Lonnie Thompson has spent the past quarter century trekking to the tops of tropical mountains, from the Andes to the Himalayas, drilling cores through their glaciers, all the way to bedrock, and carrying the ice back to his freezers at Ohio State University. His goal has been to establish the same kind of long-term record for tropical climate change that the ice cores from Greenland and Antarctica have established for those regions. But he has unintentionally become an important witness to the change going on right now—which his work suggests is unprecedented for thousands of years. Since Thompson drilled his first ice core in 1983 at the Quelccaya Ice Cap in Peru, the retreat of the largest glacier there, the Qori Kalis, has accelerated tenfold, to around sixty meters a year, and an eighty-five-acre lake has formed at the glacier's snout. Recently Thompson and his colleagues recovered remains of wetland plants that had been exposed by the receding glacier. The plants dated to around 5,100 years ago—which Thompson interprets as the last time the Quelccaya area was as warm and ice-free as it is today. He believes that most tropical glaciers are likely to disappear in the near future.

When ice melts on land, the water flows down to the sea and raises sea level. In its 2001 report, the IPCC acknowledged the dire state of the world's mountain glaciers and estimated how much sea level might rise as they continued to melt in the twenty-first century. The report put it at somewhere between one and twenty-three centimeters, or less than half an inch to just under ten inches. A more substantial contribution to sea-level rise would come, the IPCC said, from the simple physical fact that seawater expands as it warms. Mostly because of that thermal expansion, the total rise by 2100 might be anywhere from nine to eighty-eight centimeters—just under four inches to just under three feet.

A rise at the top end of that range would have a substantial impact. A report by the U.S. Environmental Protection Agency in 1991, still the most authoritative account of the matter, put the cost to the United States of a sea-level rise of one meter (about three feet three inches) at between $270 billion and $475 billion—an estimate that the authors themselves immediately characterized as "optimistically low," because it assumed there would be no further development in the coastal zone after 1991. In fact, by 2004 the population of the U.S. coastal zone had increased by another 15 percent. A one-meter rise in sea level, if we didn't protect the land, would remove an area the size of Massachusetts—seventy-seven hundred square miles—from the U.S. map, a lot of it in Florida. It would swamp large parts of Miami and smaller parts of

New York and Boston, and it would wipe out valuable beaches and barrier islands all along the Atlantic and Gulf coasts.

The property it would threaten is so valuable, in fact, that we would probably be willing to spend hundreds of billions of dollars to protect it—holding back the sea with levees, as in New Orleans, raising houses on stilts, and adding sand to the beaches and barrier islands. (In New Orleans, which is already an average of two meters below sea level, the levees would have to be raised even higher.) In preventing the sea from encroaching on developed land, we would increase the loss of undeveloped wetlands; the levees would block the seas from moving inland as other areas got flooded. Wetlands are not only havens for wildlife; they also filter pollution out of runoff from the land, and they buffer the land behind them against floods. Their loss is another huge cost considered in the EPA report.

Other countries would be affected even more by a one-meter sea-level rise than the United States. The Netherlands, for instance, 70 percent of which is already below sea level—some scientists there are already proposing that their country, after centuries of holding back the sea with dikes, give up that losing battle and adopt a new vision of itself as a nation of "hydrometropoles," or floating cities. Across the North Sea in England, much of London lies below sea level and is currently protected by the Thames Barrier, a line of huge gates that can be closed whenever a storm surge threatens to roar up the river. A one-meter sea-level rise would render that massive engineering project inadequate if not altogether obsolete. The worst impacts of the rise would, of course, be felt in the poorer and even lower-lying countries, such as Bangladesh, where 35 million people live on the coastal floodplain, or Vietnam, or the various small island nations of the Pacific, some of which, such as Tuvalu, are already sinking beneath the waves. Whether the flooding in those countries would lead, as some commentators have suggested, to a flood of refugees into the developed countries is rather hard to predict. But it is worth considering.

Still, a one-meter sea-level rise spread over a century—let alone an eighty-eight-centimeter rise, the IPCC's maximum—is not the stuff of catastrophe. It would cost us, but we could mostly adapt to it. Climate skeptics often lambaste the IPCC for being alarmist and extremist and act as if this vast web of interlinked committees were some kind of tightly run environmentalist conspiracy, instead of a motley collection of mostly well-meaning, hardheaded, disputatious, and unfashionably dressed scientists traveling on stingy expense accounts. On the issue of sea-level rise, in any case, the IPCC's 2001 report was clearly far from alarmist. If anything, it was overly reassuring.

The IPCC considered a range of scenarios for future fossil-fuel emissions, plugged those emissions into a variety of climate models displaying different sensitivities to CO_2, and obtained a range of estimates for how much Earth might warm by 2100. They then estimated how much that warming might expand the water already in the sea, and how much ice it would melt on mountaintops around the world. The result was the forecast that sea level might rise as much as eighty-eight centimeters or as little as nine. But only a tiny fraction of all ice on land is locked up on mountaintops. Ninety percent of it, around 6 million cubic miles, is in the Antarctic ice sheet, which averages more than a mile thick. Most of the rest is in the Greenland ice cap. The IPCC concluded in 2001 that not much was going to happen to those two reservoirs in the twenty-first century. The most likely outcome was that Greenland would melt a bit, raising sea level by an inch or so. But that rise would be more than offset by an increase in the amount of snow falling on Antarctica, which would thicken that ice sheet and draw sea level down by several inches. In sum, the ice sheets were actually going to protect us from sea-level rise, at least for the next century.

The IPCC's conclusion was reasonable in 2001, and it is still far from certain that it is wrong. But the news since then from Greenland and Antarctica has not been reassuring.

Even in 2001, some evidence painted a more disquieting picture. It came from the study of the last ice age and from the interglacial period that preceded it. After Broecker and Matthews dated the coral terraces on Barbados in 1967, and determined that the last interglacial period had occurred 124,000 years ago, other researchers found raised coral terraces on coasts that, unlike Barbados, were tectonically stable. That made it possible to estimate sea level during the last interglacial. The results showed it was between four and six meters—twenty feet—higher than today. The only way for sea level to have been that much higher is for there to have been less ice on Earth then than there is now.

There is no question that there was less ice on Greenland. At Dye 3 in southeastern Greenland, where today the ice is a mile and a quarter thick, there was no ice at all during the last interglacial; when the drill team hit bedrock there in 1981, the oldest ice they found dated not from the last interglacial or before, but from the ensuing glacial period, the most recent one. Apparently all the ice from the previous glacial had melted. Sediment cores confirm that Earth was warmer in the last interglacial than it is now, especially in the high northern latitudes. So does the fossil record; it shows, for instance, that the boreal forest reached as far north as the coast of the Arctic Ocean almost all around its perimeter, covering hundreds of miles of what is now tundra.

Jonathan Overpeck of the University of Arizona, Bette Otto-Bliesner of the National Center for Atmospheric Research, and their colleagues have simulated the climate of the last interglacial with the NCAR climate model, using the paleoclimatic data as constraints. Their results, reported in *Science* in 2006, suggest that average temperature in Greenland was only about 3.5 degrees Celsius, or 6.3 degrees Fahrenheit, warmer than it is today. But that was enough to raise the maximum high temperature in summer above the freezing point over the entire ice sheet—which meant that the ice was subject to at least some melting everywhere, if only for one day a year. In the simulation, the ice sheet melted back until all that was left was a steep dome in central and northern Greenland. Along with melt from the smaller ice caps in Iceland and the Canadian Arctic, that raised sea level as much as 3.4 meters, or 11 feet, higher than it is today. The researchers concluded that the rest of the sea-level difference—that is, between one and three meters—must have come from melting in Antarctica. Sediment cores from the Ross Sea suggest the ice shelf there was smaller; ice cores indicate it was warmer in Antarctica in the last interglacial. It seems there was less ice at both poles then than there is today.

That makes sense, because the Milanković cycles were pouring more intense summer sunlight onto the high northern latitudes than they are today. (Recall that Broecker's dating of the last interglacial in Barbados was one of the first bits of hard evidence in support of the Milanković theory.) On the other hand, the amount of carbon dioxide in the atmosphere was about the same in the last interglacial as in the nineteenth century, 280 parts per million. Today it is more than 380 parts per million, and rising fast. Business as usual would triple the preindustrial level by 2100. Overpeck, Otto-Bliesner, and their colleagues ran their simulation into the future, high-CO_2 world. The NCAR model is less sensitive to CO_2 than some others, but they nevertheless found that by 2100, temperatures in Greenland and Antarctica exceeded those in the last interglacial. The ice in Greenland may also melt more readily now than it did in the last interglacial, because it is being darkened by industrial soot and thus absorbs sunlight better. By 2100, the researchers concluded, a lot of melting in Greenland and Antarctica could have become inevitable, and thus "a threshold triggering many meters of sea level rise could be crossed."

That does not necessarily imply that such a large sea-level rise will have already occurred by the year 2100 (as some press reports suggested), which would be a real catastrophe. No one knows how fast the ice would go away—it could take centuries. There too, though, the evidence from the past is not altogether reassuring. In the 1980s a colleague of Broecker's at Lamont, Richard Fairbanks, followed up on the work Matthews and Broecker had done on Barbados. Instead of just dating the exposed reef terraces on the island, which originated in the last interglacial and the beginning of the last ice age, Fairbanks and his team located more recent terraces offshore—terraces that were submerged by the sea-level rise at the end of the last ice age and have not yet been tectonically lifted above the waves. By drilling into them, Fairbanks showed that sea level has risen by 120 meters, or 400 feet, in the twenty thousand years since the Last Glacial Maximum. More important, at the end of the Younger Dryas, it rose fifteen meters in a millennium, which is five feet a century—nearly twice as fast as the fastest rate in the IPCC report. During the Bölling/Alleröd, around fourteen thousand years ago, Fairbanks thought he could pinpoint a time when sea level rose twenty meters in a single century. That claim has always been controversial, but it hasn't been disproven either.

It takes a long time for a climate warming to propagate by conduction through a mile or more of ice; as we mentioned before, the news that the air over Greenland is now warmer than it was during the last ice age has yet to make it to the bottom of the Greenland ice sheet. If all we have to fear from the ice sheets is that they will melt from the surface down, like an ice-cream cone in the sun, then we don't have much to fear from them in this century—the process will be a slow and steady wasting over many centuries. It will basically be predictable. That was the assumption underlying the last IPCC report in 2001. Yet we know from the past that slow melting is not the only way to get rid of ice—sometimes gargantuan chunks of the stuff break off and slide into the sea, in a way that no glaciologist's model can yet predict. This process happened on a truly massive scale at least six times during and at the close of the last ice age: those are the Heinrich events, recognizable in Atlantic sediment cores by the distinctive debris deposited by the armadas of melting icebergs.

The source of some of those Heinrich events was a large dome of ice centered over Hudson Bay. It may have been prone to collapse because its base was below sea level and thus exposed to erosion by seawater, and because it sat on slippery, waterlogged sediments instead of solid rock. An ice sheet that is grounded below sea level is called a marine ice sheet, and during the Ice Age there were many of them. Now there is one left: the West Antarctic Ice Sheet. Some scientists have been worried about the WAIS for decades, but in its 2001 report the IPCC decided there was no immediate cause for concern. It didn't matter that glaciologists don't have a model that can explain the way ice sheets move, because the WAIS was not going to move anytime soon. "It is now widely agreed that major loss of grounded ice from the WAIS and consequent accelerated sea-level rise are very unlikely during the

21st century," the IPCC report said. The agreement was not universal then. It is even less so now.

One of the first people to worry about the West Antarctic Ice Sheet was a researcher at Ohio State named John Mercer. Mercer was a tall, skinny, shy man, a geographer not a glaciologist by training. But he spent a considerable amount of time tramping around ice fields in Antarctica, South America, and New Zealand, generally taking only one or two people with him. Sometimes, recalls Lonnie Thompson, whom Mercer introduced to the Andes, he did his fieldwork in the nude (although presumably not in Antarctica). He was once arrested for jogging nude in a Columbus park in the predawn hours. Richard Alley, who as a graduate student also knew Mercer, remembers him as "polite, but with perhaps an imperfectly concealed hint that he would be more comfortable on an outcrop than sitting in the office discussing the outcrop with me." George Denton, who encountered Mercer often in Antarctica, remembers him as a man of tremendous intuitive rather than analytical intelligence, a man who could see a landscape and "get" the processes that had produced it. It is not always easy for such a person to defend his conclusions in front of a roomful of analytical types, which glaciologists tend to be—the problem of understanding how ice sheets move, which is still far from solved today, is a physics problem. When Mercer spoke publicly, he spoke hesitantly; when he wrote, though, he was far more forceful. He did not hesitate to use the first-person singular even in professional papers, and vigorous words such as *disaster*. Neither his speaking nor his writing style was such as to increase his influence among his peers.

West Antarctica faces the Pacific; it is bounded in the east by the Antarctic Peninsula (the long finger that juts up toward South America), in the west by a deep bay called the Ross Sea, and inland by the Transantarctic Mountains, which separate it from East Antarctica. West Antarctica contains about 10 percent of the continent's ice, East Antarctica the rest. Most of the East Antarctic Ice Sheet sits on bedrock above sea level; most of the West Antarctic Ice Sheet sits on the soft seabed, in some places more than a mile and a half below sea level. Because the ice is more than two and a half miles thick in some places, it rises above sea level. Ice flows in narrow streams from the domed interior of the sheet toward the sea. The streams can move more than half a mile a year, but at least one of them hasn't moved in decades, for reasons that are mysterious. The thickness of the ice tapers as it flows, until eventually, at a place called the grounding line, it floats free of the seabed and continues as an ice shelf. The ice sheet gains mass from snow falling on its interior, and loses mass to melting, and above all to the icebergs that calve off the ice shelves at their fringes.

If the West Antarctic ice sheet were ever to disintegrate completely, sea level would rise by five meters, or more than sixteen feet. The Transantarctic Mountains, which now look out onto an ice desert that is up to 700 miles wide, would instead face a sea with several large islands. Could it ever happen? It has happened in the past. Mercer was convinced it had happened during the last interglacial—that West Antarctica was more likely than Greenland to account for the sea-level rise then, because as a marine ice sheet it was far more vulnerable to collapse. The main thing that kept it from draining into the sea, he thought, were the floating ice shelves, which acted like corks to keep the ice bottled up. If the air or sea around Antarctica were ever to warm enough to remove the ice shelves, Mercer argued, the ice sheet would collapse, fast.

In 1974, the Wordie Ice Shelf on the west coast of the Antarctic Peninsula partially broke up, scattering six hundred square kilometers of ice, a quarter of its area, into the Bellingshausen Sea. Mercer considered that a bad sign. In 1978, three years after Broecker warned that "we may be in for a climatic surprise," Mercer issued his own warning in *Nature*:

> I contend that a major disaster—a rapid 5 m[eter] rise in sea level, caused by deglaciation of West Antarctica—may be imminent or in progress after atmospheric CO_2 content has only doubled. This concentration of CO_2 will be reached within about 50 years if fossil fuel continues to be consumed at its recent accelerating rate.

It was not that no one listened to him. A few people did. Even before Mercer's paper, glaciologists were debating what controlled the flow of the WAIS, and what kept it from succumbing to gravity and flowing right out to sea. One camp thought that any particular ice stream was held in place by "local" forces—by friction at its base, and by friction with the stationary ice on either side of a stream. Mercer belonged to the other, "nonlocal" camp; he thought the whole ice sheet was held back by the distant buttressing force of the ice shelves. But his picture was more intuition than quantitative physics. Even today there is no physical model that includes all the forces acting on an ice sheet and explains, for instance, why the fast-moving ice streams exist, and why they sometimes stop.

And by the time of the IPCC and its reports, the idea of ice shelves as buttresses had fallen out of favor, for no conclusive reason. The idea that the West Antarctic Ice Sheet would remain basically stable in this century—to say nothing of the more solidly grounded Greenland ice sheet—had become orthodoxy. In his 1978 paper, however, Mercer had proposed a test of his idea. "One of the warning signs that a dangerous warming trend is underway in Antarctica," he wrote, "will be the breakup of ice shelves on both coasts of the Antarctic Peninsula, starting with the northernmost and extending gradually southward." Mercer recommended that the shelves be monitored with satellite images.

In the southern summer of 1979, the year after Mercer's paper, researchers examining Landsat images first noticed melt ponds on the surface of the Larsen Ice Shelf, on the east coast of the Antarctic Peninsula in the Weddell Sea. The Larsen Ice Shelf was fed by a series of glaciers flowing off the land, and from north to south, it was divided into three main sections, called A, B, and C.

By the 1990s, satellite pictures showed the melt ponds increasing in number and extent, which made sense: the Antarctic Peninsula was dramatically warming. In the last half century it has warmed between two and four degrees Celsius; only the Arctic has warmed as much. In January 1995, the melt ponds were more extensive than ever. That month, the Larsen A Shelf suddenly shattered and exploded, and sixteen hundred square kilometers of ice, more than six hundred square miles, drifted off into the Weddell Sea as scattered bergs.

Why had the ice shelf shattered instead of just melting gradually as temperatures rose? Ted Scambos of the University of Colorado and his colleagues proposed that the melt ponds were to blame. A floating ice shelf, like an ice sheet or glacier on land, is always scarred by crevasses, cracks formed by stresses in the flowing ice. The crevasses are mostly surface scratches—they don't penetrate right through to the bottom of the ice shelf, which is hundreds of meters thick. At a much shallower depth, the weight of all the overlying ice tends to squeeze the crack shut. But if a melt pond forms at the surface, and if it finds such a crack to drain into, and if that crack is deep enough—Scambos's colleague Christina Hulbe at Portland State University showed it didn't need to be deeper than fifteen meters—the weight of the water in the crack tends to wedge it open. The crack can then go right through the shelf.

That was the theory—and in 2002 it passed a dramatic test. In February, the Larsen B Shelf shattered in the same way as Larsen A, but scattering twice as much ice this time, around 3,275 square kilometers. Satellite images revealed that there had been extensive melt ponds in the part of the shelf that had shattered, the northern part; but the southern part, with no melt ponds, had remained intact. And just before the breakup, the pictures showed individual large melt ponds suddenly contracting in size—apparently because they were draining into cracks.

Even more important than the confirmation of the theory of how ice shelves break up, and of Mercer's prediction that they would break up, from north to south, was the confirmation of his theory of what happens next. After the Larsen B explosion, Scambos's group and another led by Eric Rignot of the Jet Propulsion Laboratory monitored the glaciers that feed into the ice shelf. The two groups used different sorts of instruments on different satellites to measure the height and speed of the glaciers. They got the same result: there was no change in the glaciers feeding into the intact part of the shelf—but the glaciers behind the disintegrated part had dramatically accelerated. Rignot's team found that three of those glaciers were moving eight times faster than they had been in 2000, two years before the breakup. This was strong evidence that Mercer's basic idea, that ice shelves buttress ice sheets, was right.

It is not the whole story, however. Melt ponds don't just accelerate the breakup of floating ice shelves; they also seem to accelerate the flow of the glaciers themselves. So far the effect has been better documented in Greenland, which is warmer than most of Antarctica, and where every summer large lakes of meltwater form on the ice sheet, especially near the coasts. In the late 1990s, Jay Zwally of NASA's Goddard Space Flight Center and his colleagues stuck a pole in the ice at a place called Swiss Camp, about thirty-five kilometers inland from the ice edge in west-central Greenland. They equipped it with a GPS receiver so they could track its movements precisely. In the winter, the pole typically moved seaward at around thirty-two centimeters a day, just over a foot. But in August of 1998 and 1999, it advanced as much as forty centimeters a day—just as meltwater lakes were forming in the vicinity, and torrents of meltwater were pouring into wide cracks called moulins. Zwally and his colleagues think the water was making it all the way through the mile-thick ice, and by lubricating and softening the underlying bed, making the whole ice sheet surge faster. The phenomenon had long been observed on mountain glaciers, but never before on an ice sheet. Though there is less evidence of it so far in Antarctica, it may have contributed to the acceleration of the glaciers on the peninsula after the breakup of Larsen B.

Another mechanism might have contributed too. A floating ice shelf and the glacier behind it can be melted not only from above, by warm air, but even more efficiently from below, by warm water. The ocean is getting warmer: just as it is soaking up a third of the CO_2 we are putting into the atmosphere, it is soaking up more than 80 percent of the heat already trapped by that increasing atmospheric CO_2. Since 1955, according to data compiled by Sydney Levitus and his colleagues at the National Oceanic and Atmospheric Administration, the world ocean has warmed by an average of nearly four-hundredths of a degree Celsius. (That doesn't sound like much, but if all that heat were transferred to the much less massive atmosphere, it would warm by forty degrees Celsius.) In the Southern Ocean, the surface waters have not warmed so much as the waters at intermediate depths of six hundred meters or so. Rignot and Stan Jacobs of Lamont-Doherty have calculated that the intermediate water just off Antarctica has warmed by 0.2 degree Celsius in recent decades.

As that warmer water laps up onto the continental shelf and under the floating ice shelves, it thins them. And it may be that the ice shelves do more than buttress the glaciers behind them, as Mercer thought; they may also insulate the glaciers from the warm seawater. Once an ice shelf disappears, that warm water, unchilled by overlying ice, can attack the glacier directly at its grounding line, the place where it meets the seabed.

That may already be happening in the Amundsen Sea, west of the Antartic Peninsula. The glaciers there have either narrow ice shelves or none at all. In 1998, Rignot reported satellite radar observations indicating that one of those glaciers, the Pine Island Glacier, was retreating rapidly—around three-quarters of a mile per year between 1992 and 1996—as the ice drained more rapidly into the sea. Then in 2004, Rignot, Robert Thomas of NASA, and their colleagues completed a far more extensive—and alarming—survey of the Amundsen Sea. Flying along the coast in a Chilean navy plane, they measured the height and thickness of six glaciers with a laser altimeter and with ice-penetrating radar. They found that all the glaciers had rapidly been thinning since the 1990s, and at an accelerating pace. The Pine Island Glacier was losing four feet a year in thickness, twice as much as had been estimated in the 1990s.

With his satellite radar Rignot documented the same disturbing trend in Greenland. In 2006 he reported that the Jakobshavn Glacier on the west coast, already known as one of the world's fastest melting, had doubled its speed in recent years. On the east coast, the Kangerdlugssuaq and Helheim glaciers had more than doubled theirs. Near the coast the Jakobshavn Glacier has been thinning by fifteen meters per year for the past decade; Kangerdlugssuaq thinned by forty meters a year between 2003 and 2005. As NASA glaciologist Robert Bindschadler has pointed out, those glaciers, like the ones in Antarctica that have been accelerating and calving icebergs the most, are the ones that have cut the deepest channels into the bedrock. Because they are deep, they have the greatest area exposed to water where they meet the sea, and so they are most vulnerable to attack.

Kangerdlussuaq alone, Rignot calculated, dumped nearly 33 billion metric tons of ice into the Greenland Sea in 2005. Enormous masses of ice like that don't slide smoothly to the sea—they lurch. In 2003 Göran Ekström of Harvard and his colleagues reported the discovery of a hitherto neglected natural phenomenon: glacial earthquakes. The vibrations triggered by a surging glacier as it shakes the rock beneath it are powerful enough, it turns out, to be detected by seismometers around the world. But because they are of a much lower frequency than those of ordinary tectonic earthquakes, they had not been noticed before. In seismic records dating from 1993 through 2005, Ekström's group traced 136 quakes, all of magnitude greater than 4.6, to the outlet glaciers of Greenland—Kangerdlussuaq alone had triggered 26. The number of quakes increased more than sixfold between 1993 and 2005, especially after 2003, right when Rignot was seeing a dramatic acceleration of the glaciers. The typical glacial earthquake, the Harvard researchers calculated, corresponds to a mass of 10 billion tons of ice sliding thirty feet or so in thirty to sixty seconds.

It ought not to surprise us, but it usually does, to learn that huge events are happening on Earth that we are totally unaware of. Huge events have been happening in both Greenland and Antarctica that were not dreamt of even by scientists deeply concerned about global warming—with the exception of a few prescient souls such as John Mercer.

What is happening to the level of the sea as all that ice falls into it? The answer is not as obvious as it sounds. To begin with, when floating ice shelves such as the Larsen—or the entire Arctic ice cap—disintegrate and melt, sea level does not significantly rise, any more than the water in a glass rises when the floating ice cubes melt. The ice was already displacing its weight in water, according to Archimedes' principle, and the meltwater displaces nearly the same amount. In the case of ice floating on the ocean, the meltwater displaces slightly more, because it is fresher and thus less dense than ambient seawater, and so sea level does in fact rise, but only slightly. When a glacier on land or frozen to the sea bottom slides into the sea, though, all that water is additional, and so sea level rises appreciably—like the level of water in a glass when you drop an ice cube into it.

But global warming does not merely increase the flow of meltwater *to* the sea; it also increases the evaporation of water *from* the sea. Climate models predict that the warmer air should carry more of that water vapor over the interior of Greenland and Antarctica, where it would fall as snow, thus counterbalancing to some extent the loss of ice along the periphery of the ice sheets. And there have in fact been reports, based on satellite altimetry measurements, that both the Greenland and East Antarctic ice sheets have been thickening in recent years. On the other hand, the best measurement yet of snowfall trends in Antarctica, based on ice cores scattered around the continent, indicates no increase in snowfall over the Antarctic continent as a whole in the past fifty years. The IPCC report in 2001 counted on such increased snowfall in predicting that the ice sheets would not add to sea level over the next century. The snows may yet come, but apparently they haven't started yet.

On the contrary, a gathering flood of evidence indicates that ice melting is already winning out and the ice sheets are adding substantially to sea level. From their satellite radar measurements, Rignot and Thomas calculated that the Amundsen Sea glaciers alone were discharging more than 230 billion tons of ice into the sea each year—around 80 billion tons more than they were catching as snow, and enough to raise the sea level by a quarter of a millimeter, or a tenth of an inch. According to Rignot the Greenland glaciers are now adding half a millimeter a year to sea level.

Meanwhile a completely different technique has yielded essentially the same result for Greenland, in the hands of two independent teams of researchers. Instead of using satellite radar to measure the height, horizontal motion, or thickness of an ice sheet, this technique uses a pair of satellites called GRACE to measure the changing volume of the ice. One satellite follows 137 miles behind the other, on the same orbit three hundred miles above Earth. As they pass over denser concentrations of mass on Earth's surface, the extra gravity of that extra mass accelerates the satellites, one after the other. That causes the distance between them to widen and then narrow again, as if they were connected by a spring. A microwave beam transmitted between the two satellites allows that distance to be measured to within less than the width of a human hair—from which the gravity field can be calculated with equivalent precision. GRACE maps the entire gravity field every thirty days, so it can detect not just spatial variations but also variations over time in the distribution of mass on Earth's surface. For instance, it can detect large quantities of ice crashing into the sea and melting.

In September 2006, Isabella Velicogna and John Wahr of the University of Colorado reported GRACE measurements of the amount of ice loss from Greenland. Like Rignot, they found that Greenland is adding half a millimeter per year to sea level now, mostly from glaciers in the south. And like Rignot too, they found that the melt rate had more than doubled recently—they said the change had happened in 2004. A team from the University of Texas that analyzed the GRACE data got virtually identical results, including the acceleration in 2004. A third team, using a different method of data analysis, found a much lower rate of ice loss—but it still amounted to 0.2 millimeter of sea level per year, and they too saw a large increase in melting in Greenland since the 1990s.

Something is happening in Greenland, and in Antarctica too—Velicogna and Wahr's estimate of ice loss from that whole continent suggests that it has been adding .4 millimeter per year to sea level. If in spite of there being no evidence of increased snowfall, the East Antarctic ice sheet is indeed thickening—the researchers couldn't tell from the GRACE data—West Antarctica is losing ice even faster.

For much of the twentieth century, tide gauges scattered around the world have recorded a sea-level rise whose global average has been calculated to be around two millimeters a year. (Locally measured sea level varies a lot because the land on which the tide gauges sit is subsiding in some places and rising in others.) Since the early 1990s, when more accurate satellite measurements became available, sea level has been rising at nearly three millimeters a year. If things were to continue at that rate, by the year 2100 sea level would have risen around a foot—not a trivial amount. It would cost lot of money to protect our coasts from that, but it would be manageable.

The worry is that sea level might start rising at a much faster rate. What the recent observations in Greenland and Antartica suggest is that, contrary to the IPCC's sanguine analysis in 2001, those ice sheets may already be contributing nearly a millimeter of the three-millimeter sea-level rise happening now—and that makes people nervous. "I had a higher comfort level a few years ago," says Robert Bindschadler, who has been coordinating a research program on the West Antarctic Ice Sheet since 1990. "Now I just don't know. These newer revelations of just how rapidly the ice sheets have changed in the last few years caught all the experts off guard."

"The IPCC said, 'We've got a hundred years before anything happens to the ice sheets,' " says Richard Alley. "It looks like the ice sheets are a hundred years ahead of schedule."

Yet on February 2, 2007, when the IPCC released the initial summary of its new, fourth assessment report, it actually *reduced* the maximum sea-level rise projected for the year 2100 from eighty-eight centimeters to fifty-nine centimeters. The IPCC scientists specifically did not take into account the recent observations of accelerated ice loss from Greenland and Antarctica—essentially because they didn't know what to make of them. "Dynamical processes related to ice flow not included in current models but suggested by recent observations could increase the vulnerability of the ice sheets to warming, increasing future sea level rise," the report said. "Understanding of these processes is limited and there is no consensus on their magnitude."

Whether you find that statement comforting depends on your tolerance for risks, probably small, of disasters that would certainly be large. We do not find the statement comforting.

In the long run, as Earth warms, sea level will inevitably rise. If we stay our present course of rising CO_2 emissions, sea level will rise a lot. In the long run, of course, we are all dead. Will our children and grandchildren be dead before sea level rises a lot? We don't know. We would like glaciologists to tell us, but glaciologists do not really understand what has been happening recently in Greenland and Antarctica. Still less do they have a model that would allow them to make a reliable prediction of how those ice sheets will move in the future. The past remains our best guide. The past shows that big sea-level changes have happened, and it suggests that they have sometimes happened fast.

During the last interglacial, as we've already seen, sea level was as much as six meters higher than it is today. During the Pliocene Epoch, 3 million years ago, when Earth was just three degrees Celsius warmer than today—about what we can expect from doubling CO_2—sea level was twenty-five to thirty-five meters higher than today; Florida did not exist. Even the last century is instructive. Stefan Rahmstorf, an oceanographer and climate modeler at the Potsdam Institute for Climate Impact Research in Germany, recently compared the data for temperature changes and sea-level changes in the twentieth century. He found, unsurprisingly, that they were closely correlated—that is, sea level rose in direct proportion to temperature. Using that simple relation and the IPCC's own range of scenarios for CO_2 emissions and temperature increase between now and 2100, Rahmstorf did his own calculation of what sea level might be in 2100. His range was substantially higher than the IPCC's—50 to 140 centimeters, instead of the 18-to-59-centimeter range in the IPCC's latest report.

And Rahmstorf's estimate too assumes that nothing substantial happens to the ice sheets—that we are not witnessing the beginning of a sequence of events in Greenland and Antarctica that is outside our historical experience. But we may be witnessing just such a change. In Greenland, most of the ice loss so far has occurred from glaciers along the southeastern and southwestern coasts, but Rignot's measurements suggested that the glaciers up north are beginning to accelerate as well, as the warming creeps north. Meanwhile the area of the ice sheet that is exposed to at least some melting every year, once a thin band along the coast, is steadily creeping inland. Meltwater lakes are becoming bigger and more frequent. As they move inland, they may leak through to the bottom of the ice sheet and lubricate the bed there, as they have already been shown to do near the coasts. Vast sections of the ice sheet might then become far more mobile than they are today. Ice loss might then abruptly increase. That is the worry anyway.

In Antarctica, the air has not yet warmed much, with the exception of the Antarctic Peninsula. One reason seems to be the famous hole in the stratospheric ozone layer, which is centered over Antartica. Ozone warms the stratosphere by absorbing solar radiation; ozone depletion, conversely, cools the stratosphere, and the cooling, climate modelers at the Goddard Institute of Space Studies in New York have found, intensifies the ring of westerly surface winds that isolate Antarctica from warmer air to the north. Under the 1987 Montreal Protocol—one of the great environmental success stories—the nations of the world have agreed to phase out their use of ozone-destroying chlorofluorocarbons. Thus by the second half of this century, the ozone layer should have recovered—and warming in Antarctica, climate models suggest, should have caught up with that of the rest of the world, or perhaps surpassed it. It is reasonable to fear that the loss of ice will accelerate then.

Another reason the air over Antarctica has not yet warmed as much as the air in the Arctic is that there is more ocean in the southern hemisphere to take up the heat. As the ocean continues to warm, the melting of ice shelves and glaciers in Antarctica will accelerate. Sea-level rise itself is expected to accelerate the melting. In his thirty years of Antarctic fieldwork, George Denton found that the West Antarctic Ice Sheet was somewhat larger at the Last Glacial Maximum; the Ross Sea, for instance, the large embayment that is now half covered with a floating ice shelf, was then completely filled with grounded ice, which extended outside the mouth of the bay to the continental shelf. Since the LGM the ice sheet has sagged back to its current position. Some of that retreat has been driven directly by warmer temperatures in Antarctica, but even more, Denton thinks, has been driven by the rise in sea level produced by the warming in the northern hemisphere. As the Laurentide and Scandinavian ice sheets disintegrated, the rising sea ate away the grounded ice in the Ross and Weddell seas, causing more ice to flow into the sea from the interior of West Antarctica, and raising sea level still further.

The great fear is that man-made global warming may now be pushing that process to some kind of terminal threshold—that the West Antarctic Ice Sheet is on the verge of collapse. As warmer water and a rising sea attack the base of the ice sheet's outlet glaciers, a positive feedback could set in that would accelerate the process dramatically. The ice sheet is farthest below sea level toward the center, where it is thickest and its great weight presses down most on the bedrock. Thus as ice on its perimeter melts, and the edge of the ice sheet retreats inward, an ever greater surface is exposed to the warm seawater, and so the melting should go ever faster. In their 2004 aerial survey of the Amundsen Sea glaciers, Thomas and Rignot discovered not only that the glaciers were shedding ice more rapidly, but also that they were much deeper at the coast than had been thought—an average of four hundred meters deeper, or more than thirteen hundred feet. In some places the bedrock was as much as a mile and a quarter below sea level. Those submarine channels seem to connect to a large submarine basin hundreds of miles inland. As the researchers put it, the glaciers thus provide "exit routes for ice from further inland if ice-sheet collapse is underway."

Back in the 1970s, Denton's Maine colleague Terry Hughes, who shared Mercer's views on the buttressing effects of ice shelves, suggested that the Amundsen Sea might be the "weak underbelly" of the West Antarctic Ice Sheet: the channel through which the whole ice sheet might collapse into the sea. Hughes's hypothesis fell out of favor along with Mercer's. It is taken more seriously now. If the entire West Antarctic Ice Sheet were to disintegrate, sea level would rise more than sixteen feet. The process might take centuries, but it might go much faster. Certainly it will go faster the more CO_2 we put into the atmosphere, and the warmer we allow the planet to get.

ELEVEN

Megadroughts of the Past

The streams and rivers that rush down off the eastern flank of the Sierra Nevada, in California, don't empty into the sea; they empty into the desert. On the western edge of the Great Basin they form a series of lakes, of which the largest is Pyramid Lake. In 1955, on his way from Los Angeles to Pyramid Lake with Phil Orr, Broecker passed what was left of two of the others. Owens Lake, which had once covered more than a hundred square miles and been up to fifty feet deep, had long since been reduced to a salt flat; after Los Angeles had opened its new aqueduct in 1913 and had started diverting the Owens River into that 233-mile-long channel, the lake had dried up in a little more than a decade. "Whoever brings the water brings the people," had been the famous dictum of William Mulholland, chief engineer of the Los Angeles Department of Water and Power—but by the 1930s, so many people had come that the LADWP was having to extend the aqueduct farther north along the Sierra. To capture the streams that feed Mono Lake, the LADWP drilled an eleven-mile tunnel right through the Mono Craters, a volcanic system that last erupted only six hundred years ago. By the time Broecker and Orr passed Mono Lake in 1955, its level had dropped fourteen feet.

The cause was unprecedented, but not the effect; Mono Lake has been up and down many times in the past. As Broecker learned at Pyramid Lake that summer, terminal lakes are like multicentury rain gauges. Until humans started diverting their supply rivers, they had no outlet except the sky—water escaped only by evaporation. As the rains came and went and the temperature rose and fell, the water level in the lakes went up and down, tracking the difference between precipitation and evaporation. At the end of the last ice age, around fifteen thousand years ago, when Lake Lahontan desiccated except for Pyramid Lake and a few other remnants, it lost 90 percent of its surface area; its level dropped more than five hundred feet, and the pine forests that once graced its shores became sagebrush desert. Meanwhile Mono Lake dropped nearly seven hundred feet. The rains and snows that were no longer falling on the Sierra Nevada and the Great Basin were now falling somewhere else—in the African Rift Valley, for instance, where Lake Victoria, which had dried up during the Ice Age, was born again.

When Earth gets warmer or colder, it is not just the temperature that changes, and not just the sea level. The whole global water cycle changes too, shifting the distribution of rain—and with it the boundaries between places that are eminently livable for people, and others that are not so. The changes that result from our current global warming may or may not turn out to be as large as the ones at the close of the Ice Age. The same climate models that predict a warming of 1.5 to 4.5 degrees Celsius if we double atmospheric CO_2 estimate that the planet warmed by around 4 to 7 degrees Celsius after the Ice Age. Predicting the global average temperature is hard enough, but predicting what people really care about, which is how hot it will get and especially how much water will be available in a specific region, is much harder. The models are just getting to the point of being able to do that. The problem is that adding or taking away an inch of rain makes a big difference in a place like the Great Basin, which typically gets seven or eight inches a year—but it is a small fluctuation in the global water cycle.

The fluxes in that cycle are staggering. In 2007 Kevin Trenberth and his colleagues at the National Center for Atmospheric Research in Boulder, Colorado, published new estimates of them. Every year, they calculated, 473 trillion tons of water evaporate from the ocean; of that, 40 trillion tons blow over the land, joining 73 trillion tons that have evaporated there to make 113 trillion tons of rain and snow. That would be around thirty inches of rain spread evenly over all the land on the planet.

But the rain is not spread evenly, because of two crucial facts about Earth: it rotates, and the equator gets more sunlight than the poles. Rain or snow fall only where air rises and cools enough for the scattered water molecules in it to condense into raindrops or snowflakes. Near the equator, where the sun shines straight down and heats the planet's surface most intensely, warm air laden with water vapor rises and sheds its water; under those tropical low-pressure zones lie the rain forests. At altitudes of six miles and up, the dried-out air spreads north and south and cools some more. In the subtropics, at latitudes of around thirty degrees, it subsides back to the surface. Under those high-pressure zones of dry, subsiding air lie the great deserts.

As the subsiding air nears the surface, it diverges, with some flowing equatorward and some poleward. The air flowing toward the equator is deflected to the west by Earth's rotation, and it becomes the trade winds. Near the equator, the trade winds from north and south converge, completing the circulation loop—that Intertropical Convergence Zone, as it's called, is the site of the tropical rain belt. The air flowing toward the poles in each hemisphere, meanwhile, joins a different circulation loop; as it is deflected to the east by Earth's rotation, it enters the eastward-moving storm systems of the midlatitudes. Both the equatorward flow and the poleward flow carry water that has evaporated from the surface in the subtropics. In other words, the subtropics are always exporting water to the tropics and to the midlatitude storm systems, via these two great interlocking circulation cells—atmospheric conveyor belts, you might call them.

Climate modelers are far from having worked out in detail what will happen as we monkey with this huge and intricate heat engine. But they are confident of a few things. Just as ice will melt more readily on a warmer planet, raising the level of water in the sea, liquid water will evaporate more readily, raising the amount of water vapor in the atmosphere. Right now the atmosphere contains about 12.7 trillion tons of water, which sounds like a lot but isn't—it's a little more than nine days' worth of evaporation, just enough to form an inch of rain over the entire surface of Earth. As the atmosphere warms, the water vapor molecules in it will have a

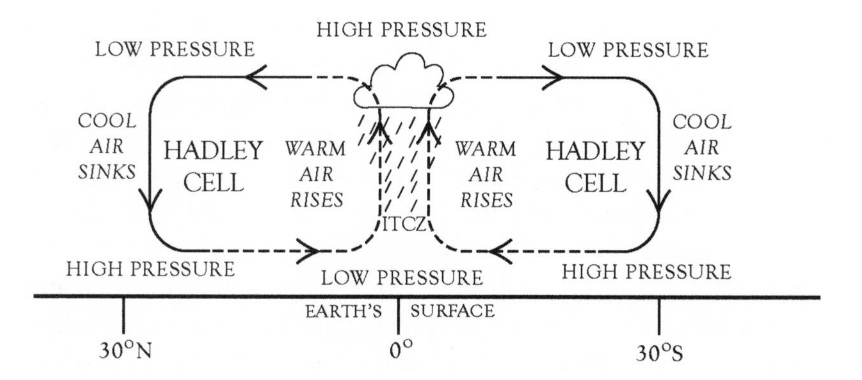

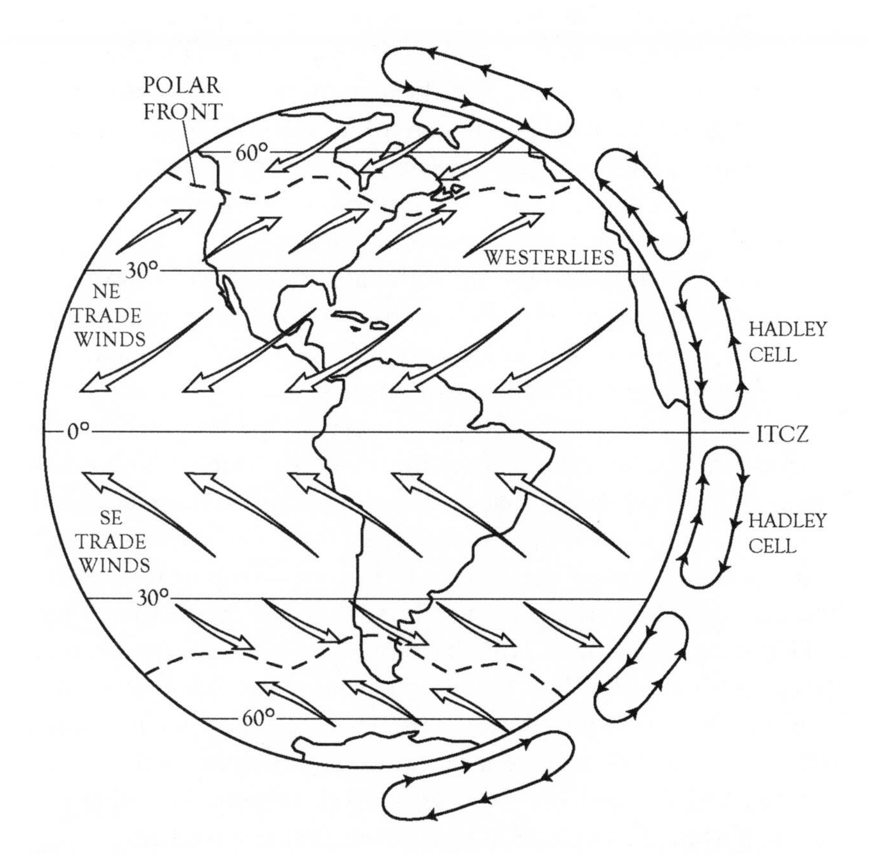

Because Earth gets more solar heat at the equator than at the poles, Hadley cells form in the atmosphere. Earth's rotation gives rise to winds from the west in the midlatitudes and to easterly trade winds that meet near the equator in the Intertropical Convergence Zone (ITCZ).

harder time sticking together to form raindrops, and so they will stay aloft longer. Thermodynamical principles and model calculations indicate that the amount of vapor in the atmosphere will go up by 7 percent for each degree Celsius of temperature rise. Evaporation and precipitation—the flows into and out of that larger atmospheric reservoir—will increase by a lesser (and less certain) amount, perhaps around 2 percent per degree Celsius. But because the amount of rain that falls at any given place and time depends on how much water vapor is available, more of the rain may fall in intense bursts. There is some evidence that is happening already.

Where will the rain fall? On average, as Isaac Held of NOAA has pointed out, the wet zones should get wetter and the dry zones drier as the planet warms. To the extent that the winds don't change much, the atmospheric conveyor belts will keep functioning as they have—but now, because more water is evaporating, they will be shipping more water vapor out of the dry subtropics and piling it up in the wet tropics and midlatitudes.

That's not the end of the story, though, because global warming will certainly change the winds to some extent. The presence of more water vapor in the atmosphere will itself change the circulation. As the vapor condenses to rain or snow, it releases a lot of heat, and that heat helps drive the circulation. One change the models predict, though atmospheric physicists don't fully understand why it happens, is that the tropical circulation loops, called Hadley cells, should expand toward the poles as the planet warms—which means the regions of dry, subsiding air that create the subtropical deserts will expand. As the subtropics expand, the midlatitude storm tracks will be pushed to higher latitudes too. None of that bodes well for the American Southwest. It is arid now, not because it lies in the subtropics—Pyramid Lake is at forty degrees north—but because it lies in the rain shadow of the Sierra Nevada and Pacific Coast Ranges. Global warming may give the region a whole new reason to be a desert, drying it further.

Finally, there is one change to the water cycle that you don't need a climate model to predict. As the planet gets warmer, some precipitation that now falls as snow will fall as rain, and the snow that does fall will melt faster. Those changes are already happening in the Sierra Nevada and elsewhere in the West. Philip Mote and his colleagues at the University of Washington analyzed measurements of the snowpack remaining on April 1 at more than eight hundred sites in the mountains of the West. They found the snowpack had thinned significantly in the second half of the twentieth century in most places, especially at lower elevations. It had even thinned in places where the amount of winter precipitation had increased—either because more of that precipitation was rain rather than snow, or because more of the snow had melted by April 1.

The West gets most of its precipitation in winter, and its demand for water is greatest in the summer. California in particular is dependent on its artificial reservoirs to stockpile the water and also to protect it from the huge floods that its volatile climate is capable of generating. The most important single reservoir, however, is the Sierra Nevada snowpack itself, which typically contains half as much water as all the built reservoirs in the state. Winter rain, or winter snow that melts too early, doesn't add to that reservoir; the water just flows to the sea at a time when the state doesn't need it and when the built reservoirs are full. Dan Cayan of the Scripps Institution of Oceanography and his colleagues have found that the time of peak spring runoff now arrives one to four weeks earlier in the Sierra Nevada (and in other Western mountains) than it did in the 1940s. Thus as the state warms over the coming century, even if it continues to receive the same total amount of precipitation, it will face a serious problem. According to modeling studies done by Cayan and other researchers, California could be facing a severe threat to its water supply by 2100—a loss of 30 to 70 percent of the Sierra snowpack, depending on whether we triple or merely double atmospheric CO_2.

The American West is not the only region of the world at risk of drought and threats to its water supply; it is just the richest. A sixth of the world's population depends on glaciers or snowpack for water—large parts of India and China, for instance, get their water from rivers fed by shrinking Himalayan glaciers. In its latest report in 2007, the IPCC projected with "high confidence" that the frequency of drought and the area affected would increase worldwide in the twenty-first century. Africans and Asians will no doubt suffer more from such disruptions to the water cycle than Americans. The suffering caused, for instance, by drought in the African Sahel region during the 1970s and 1980s created images that are burned into many of our minds. But Americans are the ones who must lead the way if the world is to prevent the worst impacts of global warming. So the American West is worthy of special attention.

In the West, Americans conquered the desert with engineering—with marvels such as the Los Angeles Aqueduct, which delivers pure mountain water to the city by gravity alone, even generating electricity along the way at a dozen hydroelectric plants; or such as Hoover and Glen Canyon dams, which harness the waters of the Colorado River Basin and make them available to seven states. We built that civilization in a particular climate, which is now changing fast. Climate models, imperfect as they still are, are one way to gauge what the future holds for that civilization. Another way is to look at the climate changes that have happened in recent centuries—to look closely, for instance, at the record stored in the Great Basin lakes. That record turns out to be deeply disquieting.

In the summer of 1979, a young graduate student from Berkeley named Scott Stine visited Mono Lake. Since Broecker's visit in 1955, the water level had dropped another thirty feet—making forty-four feet in all since Los Angeles completed the tunnel under the Mono Craters. Bumper stickers reading "Save Mono Lake" had begun appearing around the state. Stine stood with the leader of that campaign, an ornithologist named David Gaines, on the rim of Panum Crater, south of the lake. The two men gazed out at the towers of white tufa that had been exposed as the water level sank, and at the land bridge that was just then emerging between the north shore and a small volcanic cone called Negit Island. The main nesting colony of the California gull was on Negit, and Gaines's fear that coyotes would soon destroy the colony, now that they could get to it over the land bridge, was being realized. Bad as the situation looked to him, however, Gaines told Stine that the water level had been just as low before—in 1857, when European settlers first surveyed the lake, well before they had diverted any of its water. "I told him, 'That's impossible,' " Stine says. "There would have to have been this phenomenal drought."

Twenty-five years had passed since that encounter, and as Stine recalled it, he was standing at the mouth of Lee Vining Creek, one of the four creeks that the LADWP had captured in 1941. He is a tall, striking man, with a passing resemblance to Burt Lancaster and sandy hair that is now graying at the temples. Rigorous personal habits—running five miles a day, not letting food pass his lips before 3 p.m., and consuming large quantities of green tea from a thermos that never leaves his side—have kept him in the kind of shape that, along with long legs, force him to slow his pace when he has a visitor. This latest visit to Mono Lake was taking place on a beautiful morning in October 2004. The clear desert air was softly vibrating with the rhythmic squeak, like an unoiled porch swing, of eared grebes; every fall, on their way to South America, millions of those birds descend on the lake to fatten up on the trillions of tiny brine shrimp that cloud its saline waters. To the west, in the rounded cirques just under the high peaks of the Sierra, the morning sun was falling on small patches of snow.

The Sierra and much of the West were then in the fifth year of one of the worst droughts of the past century, during which the water level in the lake had dropped by four feet. But it was still six feet higher than it had been when Stine saw it in 1979. That year Gaines and other environmentalists had sued Los Angeles, claiming the city had violated a public trust by taking so much water from Mono Lake. In 1983, in a landmark decision, the California Supreme Court had accepted that argument. After another decade of litigation, the State Water Board had directed the city to restore twenty feet of water to the lake, less than half of what it had taken. Low-flow toilets had become common (if not popular) in the City of Angels, and by the time the drought hit in 1999, Mono Lake had risen by ten feet.

Through fifteen years of legal wrangling, Stine had played a crucial supporting role, as a witness for the plaintiffs. The conversation with Gaines had given him his dissertation topic; he had made himself the leading expert on past fluctuations of the lake. In his first scientific paper, he had shown that the lake level had not in fact been as low in the 1850s as it was in the 1970s—that estimate had come from a Los Angeles employee, who perhaps had a bias toward making the recent drop in Mono Lake seem unexceptional. Confronted with Stine's evidence, the city had been forced to admit in court that it had taken the lake more than thirty feet lower than its level at the time of European settlement. Meanwhile, however, Stine was discovering something even more important, and under the circumstances it was a little ironic. There really had been phenomenal droughts at Mono Lake. It's just that they had happened centuries earlier.

There was a scientific upside, Stine realized soon after his conversation with Gaines, to the city's drawdown of the lake: it had exposed a lot of history. As the creeks feeding the lake followed it down, they cut deep incisions into sediments they had deposited in their deltas at earlier periods. Standing for hours in those creeks, his sneakers sodden with icy water, young Stine found he could read the history of lake fluctuations in the stratigraphy. He could see the ripples left by waves as they neared the ancient shore, and if the creek had deposited a pinecone or some other bit of organic debris in the sediment, he could radiocarbon-date that and thus establish the age as well as the height of the shoreline. Layers of volcanic ash erupted from the Mono Craters provided other benchmarks. Panum Crater, for instance, had erupted six hundred years ago and sent an avalanche of volcanic ash tumbling into the lake; sometime after that the lake had risen to a level more than seventy feet higher than it is today and had cut a terrace in the volcanic deposit. By following that terrace around to a creek and dating pinecones in the sediment just above the terrace, Stine had established that the highstand had happened three hundred years ago, during the Little Ice Age. During that period all the cirques in the eastern Sierra, now mostly dry, were filled with active glaciers, and torrents of meltwater swelled the lake.

But the most spectacular evidence Stine found was of the lake's past lowstands. The evidence is most apparent around the mouth of Lee Vining Creek, in the broad swath of shoreline exposed by the recent recession of the lake. Hiding among the rabbitbrush, a hardy gray-green shrub that has colonized the virgin terrain, are the gnarly stumps of dead trees and shrubs. The stumps—cottonwood and Jeffrey pine and rabbitbrush and sagebrush—are rooted in place and encased in thick shells of hard, whitish gray tufa, which gives them an ancient, petrified air. Until Los Angeles intervened, the stumps were submerged in the lake, with calcium carbonate—the stuff of tufa—slowly precipitating onto them. But when the trees and shrubs were alive, they grew in the open air, on ground that must have been as dry as it is today. Stine found the stumps right down to the present shoreline—and beyond. "Even when the lake was at its lowest, all Los Angeles–induced, I could still go out into the water and find rooted shrub stumps," he says. "So that is awfully compelling testimony that nature has taken Mono Lake lower than Los Angeles took it. Which speaks to the severity of the drought that must have existed."

By radiocarbon-dating the outermost wood of each stump, Stine determined when the trees had died. The stumps belonged to two distinct generations, indicating there had been two distinct droughts. Eight of the stumps had died within a decade or two of AD 1110. The other nine had died around AD 1350. It was not the droughts that killed the trees; the droughts allowed them to live, by exposing the ground they germinated and took root in, which was close enough to the creek to give them enough water. When the droughts ended, however, the rising water of the lake drowned and killed the trees, all the members of a generation at about the same time. Those droughts must have lasted for at least as long as the trees grew. Most of the stumps at Mono Lake had some fifty annual rings in them.

But as Stine learned once he had widened his investigations to other sites in and around the Sierra, the medieval droughts lasted much longer than fifty years. North of Mono Lake, in the narrow canyon of the West Walker River, Stine discovered dozens of Jeffrey pine stumps that had been drowned by the river—some were protruding from the banks, some were in the center of the current. The canyon is so narrow that the river could not have shifted laterally enough to allow the trees to grow in those spots. But in periods of drought, with runoff from the Sierra sharply diminished, it apparently shrank to a trickle. Stine radiocarbon-dated some of the stumps: their death dates fell into the same two generations he had discovered at Mono, indicating the trees were the seed of the same two droughts. But some of these pines were more than two feet across and had more than two hundred rings in them. Those trees belonged to the first generation; the second generation included stumps with as many as one hundred and forty rings.

As Stine collected more evidence, from his own fieldwork and that of others, a startling picture began to emerge. The first medieval drought had set in, at a time when conditions were already drier than today, in around AD 900. Mono Lake fell to an elevation of 6,368 feet above sea level, a few feet below its 1980s nadir—and more than fifty feet below the level it would occupy today if none of its water had been diverted. During that same drought, Owens Lake nearly dried up. Walking around the center of the Owens salt flat, with a local guide to steer him away from quicksand, Stine found the rooted stumps of shrubs, one of which he dated to AD 1020. He also found stone tools, indicating that Native Americans had walked on the lake bed in that same epoch.

Around AD 1110, the first drought suddenly ended—with a bang. For a half century or so, the climate in the Sierra, having been substantially drier than today, became substantially wetter. Mono Lake rose more than sixty feet to 6,432 feet, a level it has never seen since Europeans first saw it. And then, by no later than 1200, the drought had returned. The lake dropped nearly fifty feet again. That second drought lasted a century and a half, until around 1350.

Two lessons emerge from Stine's chronology. One is that the climate in California is capable of changing abruptly—and presumably, although the evidence is less conclusive, it could go from wet to dry as rapidly as it went from dry to wet in the twelfth century. The second lesson is that those shifts were extreme by historical standards. *Strident* is the word Stine uses to describe the exceptional harshness of the medieval droughts. Just how harsh were they? To produce the drop in Mono Lake and Owens Lake during the first drought, according to Stine's calculations, precipitation must have fallen to around 60 percent or less of its twentieth-century average. There have been drier years—but the first medieval drought lasted two centuries. Really the word *drought* understates the phenomenon; researchers have taken to referring to the medieval events as *megadroughts*.

And it may well have been harsher even than the data from Mono and Owens lakes suggest. In 1987, Stine drove into the mountains above Mono to Tenaya Lake, one of the jewels of Yosemite National Park. It was late fall, and the lake was at its annual low. Driving by in previous years, he had noticed the tops of a dozen large tree trunks sticking out of the water. Now he took his canoe and paddled out to the trees in the center. He sounded the depth of the water with a hundred-foot measuring tape; it was seventy feet deep. He cut samples from the outermost wood of two of the trees and later had them radiocarbon-dated—one of them in Broecker's lab at Lamont, where he had started work that year as a postdoc. That tree had been drowned at the end of the first drought. The other tree died at the end of the second drought. The dates matched the Mono dates. But the precipitation estimates they they implied, which Stine confirmed at other mountain lakes in the Sierra, did not match the ones from Mono.

In October 2004, at the end of a long day spent tramping around Mono Lake and the West Walker River, Stine paused to reflect on the significance of those data. The last light was flooding over the crest of the Sierra; the thermos of green tea was now empty. In none of the five years of severe drought that had just hit the West, Stine said—indeed, in only two years since measurements began in the nineteenth century, 1976 and 1977—had precipitation in the Sierra fallen to the level needed to explain the medieval drop in Tenaya and the other high-altitude lakes. According to Stine, the reality that California's climate has such variability in it has still not sunk in with public officials.

"Here in California, we have built the most phenomenal urban and agricultural infrastructure of any place in the entire world," he says. "And we've done it since gold was discovered in 1848. The infrastructure is based on water, on storing water, on diverting water—I mean, most of the state is semiarid. And we've built it on the assumption that climate will continue to be as it has been since the gold rush.

"But at Mono Lake and elsewhere, we can document that the

past one hundred and fifty years have been aberrantly wet—some of the most aberrant precipitation conditions that we've seen in ten thousand years. So we've built this system, in a sense, on a chimera. The climate record shows that California is subject to extreme and persistent droughts, which would just throw the whole system and the whole infrastructure for a loop.

"Now, fifty or sixty percent of normal precipitation—that's a strident drought. If it were to hit California and persist, year after year after year, there would be changes. But when I get up into those montane lakes—I can't get those lakes down to their medieval levels with fifty percent precipitation. It requires a far, far more strident drought than that. They've always scared me—they're telling us that precipitation gets down to twenty-five percent of normal and stabilizes at that for a long period of time. And to tell you the honest-to-goodness truth, I've always had a hard time believing that. The evidence is there, it's slapping me in the face. But, Jesus Christ, what a drought! Maybe I don't want to believe it."

There is no doubt about the reality of the medieval megadroughts; many different kinds of evidence back up what Stine has found. At Pyramid Lake, which is fed by the Truckee River with water from the Sierra Nevada, the water level is now nearly sixty feet lower than it was in 1905, when the Bureau of Reclamation started diverting some of the Truckee into an irrigation canal. Four-wing saltbush, a drought-loving shrub with thick gray leaves, has colonized the exposed shorelines, and it is replacing sagebrush in the drier landscape around the lake. By analyzing the pollen content of sediment cores from the lake, Scott Mensing of the University of Nevada and his colleagues have shown that the sagebrush-to-saltbush ratio is an indicator of past drought—including a seventy-five-year one in the thirteenth century that corresponds to the first half of Stine's second megadrought.

Nor were the medieval droughts confined to the Sierra and its watershed—far from it. While working out of Broecker's lab, Stine found stumps dating from the first drought in three different lakes in Patagonia, near the southern tip of South America. Back in North America, other workers have found evidence similar to Stine's in the Great Plains and in the northern Rockies. In Jenny Lake, in Grand Teton National Park, a stand of adult trees is submerged in eighty to a hundred feet of water. The outer ring of one of them has been dated to AD 1350—the same year the waters rose at Mono Lake, at the end of the second medieval megadrought. One of the Jenny Lake trees still has a medieval raptor nest in it, suggesting that the waters there rose rather rapidly.

But the most extensive and best-dated evidence for medieval megadroughts in western North America doesn't come from submerged tree stumps; it comes from trees in the open air. Tree rings are highly sensitive indicators of droughts, and unlike radiocarbon dating, they allow past droughts to be dated to a specific calendar year. A number of environmental factors influence whether a tree has a good year and lays down a wide new ring of wood, or a bad year that results in a narrow ring. But in most places—and especially in the arid American West—the single most important variable is water availability. The science of dendroclimatology, the study of past climates using tree rings, was pioneered at the University of Arizona in Tucson and has spread from there around the world. An Arizona PhD from New Jersey named Ed Cook runs the tree-ring laboratory at Lamont-Doherty. Cook and his colleagues have used tree rings to create a drought history for North America that in some places goes back as far as AD 800.

They drew on the labors of many scientists who had spent many seasons in the field, drilling and extracting pencil-thin cores from ancient trees, from the bark to the pith; and then many hours in the lab, cross-dating the records from twenty to thirty trees at each site and splicing together the records to make a single long chronology. Cook's team collected 835 of those chronologies from all over North America and used them to estimate the long-term fluctuations of something called the Palmer Drought Severity Index. The Palmer index is a measure of soil moisture; climatologists calculate it today from thermometer and rain-gauge readings. But those instrument data are only available for the past century. To go deeper into the past, Cook derived a mathematical model correlating the Palmer index with the nearest available tree-ring chronologies at each of 286 evenly spaced points on a grid covering North America. He fine-tuned the model until it could reproduce the twentieth-century instrument data, then turned it loose on the past millennium. That gave him a drought history at each of the 286 points. Finally he looked at the West as a whole and calculated how much of it was in at least moderate drought in any given year.

In the fall of 2004, when Cook's results appeared in *Science*, the West was nearing the end of its fifth straight year of extreme drought. Lake Powell, the huge Colorado River reservoir created by the Glen Canyon Dam—its main purpose is to ensure there will be water left for the downstream states even in drought years—was a little over a third full, after having been nearly to the brim in 1999. The key graph in the *Science* paper put that situation in historical context. It showed that the first quarter of this century, when the states bordering the Colorado River were signing a compact to divide up its water, and Los Angeles was turning its thirsty eyes toward the Sierra Nevada, was the wettest period in the past millennium—that is, the area of the West that was subject to drought was at its lowest during that period. Not only did that result jibe with Stine's results from the Sierra, but also it reinforced a finding made three decades earlier by Cook's colleague at Lamont, Gordon Jacoby, and by Charles Stockton of the Arizona lab. From tree-ring data, Stockton and Jacoby had calculated the long-term average flow of the Colorado River at Lee Ferry, Arizona, just downstream from Lake Powell. They found it was around 13.5 million acre-feet per year. (An acre-foot is a volume of water one foot deep and one acre in extent.) The men who negotiated the Colorado River compact in 1922, however, had assumed, based on a stream gauge installed at Lee Ferry the year before, that they had 17 million acre-feet per year to work with. In other words, they had divided up around 25 percent more pie than really existed.

The four wettest periods on Cook's graph all occurred after AD 1300—and the four driest ones all occurred between AD 900 and 1300, coinciding with Stine's droughts in the Sierra. During those four dry centuries, an average of 42 percent of the West, mostly in the Southwest, was in drought, whereas the average for the twentieth century was 30 percent. During one of the four megadroughts, in the twelfth century, the drought area climbed to 50 percent and stayed there for decades. The decade of the 1150s was essentially ten straight years of drought as severe as that of 2002, when the recent drought was at its worst.

When Cook submitted the paper to *Science*, he used the word *disastrous* to describe the effects such a long-lasting and widespread drought would have were it to recur today. His editor asked him to remove the word, on the grounds that it was too subjective. Cook saw his point and chose a different way of putting things:

> Compared to the earlier "megadroughts" . . . the current drought does not stand out as an extreme event, because it has not yet lasted nearly as long. This finding shows that the West can experience far more severe droughts than any found in the 20th century instrumental climate record, including the current one.

In a later paper, for a different journal, Cook described the medieval droughts simply as "scary."

"It's my opinion that the temperature change in global warming is trivial compared to the impact on the hydrological cycle that's beginning to emerge," Cook says. "People can adapt to warmer temperatures—particularly the more technologically advanced countries—with air-conditioning. But if all of a sudden water is a problem, that's going to have a much more direct impact on people's lives.

"Water is everything to the West. In a way, nothing else matters. You take water away, and the semiarid West just depopulates." It is William Mulholland's dictum—"Whoever brings the water brings the people"—in reverse. Global warming may give us the chance to see whether it cuts both ways.

TWELVE

The Drying of the Future

At the height of the recent drought, in January 2003, another paper in *Science* made a big noise in climate-science circles. Martin Hoerling and Arun Kumar of the National Oceanic and Atmospheric Administration began by noting that the drought, far from being confined to the American West, was being felt severely around the world. Across a broad swath of the midlatitudes in both hemispheres, annual precipitation levels during the four years from June 1998 to May 2002 had been substantially below the long-term average, in some places as much as 50 percent below. In the northern hemisphere the band of dryness stretched right across the United States and the Atlantic to southern Europe and the Mediterranean, and on into Southwest and Central Asia. The common trend suggested a common cause, and Hoerling and Kumar located one: a persistent pattern of sea-surface temperatures in the tropical Pacific and Indian oceans. They called their paper "The Perfect Ocean for Drought." Colorful language is permissible sometimes even in *Science*.

The perfect ocean for midlatitude drought, Hoerling and Kumar found, is one in which the tropical Pacific is in a cool phase called La Niña. La Niña is the sister and opposite of the somewhat more notorious Niño. The alternation between the two, every three to seven years or so, is driven by an atmospheric pressure seesaw called the Southern Oscillation. In the normal state of affairs, high pressure in the eastern tropical Pacific drives the trade winds along the equator toward a low-pressure zone in the western Pacific; the trade winds in turn push warm surface water toward the west, allowing cold, nutrient-rich water to well up from the deep along the coast of South America. In La Niña, this normal pattern is amplified: the east-west pressure gradient increases, strengthening the trade winds. That deepens the pool of warm surface water in the western Pacific and Indian oceans and allows a tongue of cold water to stretch along the equator, from Ecuador and Peru way out into the central Pacific. In El Niño, the atmospheric seesaw tilts the other way, the trade winds weaken or even reverse, and the pool of warm water they've been propping up in the western Pacific comes sloshing back toward South America. Which end of the seesaw you find more fearsome depends on where you sit; each end of it brings rain in some places and drought in others. Since the 1980s, Americans have become familiar with the notion that El Niño brings rain to the Southwest—and sometimes torrential rains and mud slides to California in particular—whereas La Niña is even more strongly associated with drought in the region.

In the Niña that persisted from 1998 into 2002, the water off New Guinea and the Philippines was one degree Celsius warmer than the long-term average for the region, while the central equatorial Pacific was one degree cooler. At the same time, something equally unusual was happening in the atmosphere outside the tropics. Almost uninterrupted belts of high pressure wrapped around the planet at around forty degrees north and south; in the northern hemisphere, the belt consisted of three high-pressure systems centered on the central United States, Asia, and the North Pacific. The subsiding air under those highs caused the droughts—and the highs themselves, Hoerling and Kumar found, were caused by the unusual sea-surface temperatures in the equatorial Pacific.

If the cool tongue in the central Pacific was anomalous, what was truly exceptional—unprecedented in the twentieth century—was the intensity of the western warm pool, and that it extended so far west into the Indian Ocean, south of India. The warm pool pulled the region of tropical rainfall west with it, and the effects of that rippled throughout the atmosphere. Hoerling and Kumar simulated the atmosphere with three different computer models. They ran the models fifty separate times, starting them with different initial weather conditions, then averaged the results of this ensemble to filter out the noise of weather. They found they could only reproduce the high-pressure belt and the droughts when they "forced" the atmosphere both with that exceptional warm pool and with the cold tongue in the central Pacific. "It is thus more than figurative, although not definitive," they wrote, "to claim that this ocean was 'perfect' for drought."

That paper found a receptive audience at Lamont. In the 1980s, Stephen Zebiak and Mark Cane of Lamont had developed the first computer model capable of forecasting with reasonable accuracy the state of the El Niño/Southern Oscillation (ENSO) cycle. As a result of that work, it had for the first time become possible to make seasonal weather forecasts in the United States and elsewhere that were better than the *Farmer's Almanac*. In 1996, a graduate student in Cane's group named Amy Clement had used that model to do an interesting experiment: she allowed more energy to radiate down onto the tropical Pacific—as if the sun had gotten brighter or the atmospheric greenhouse thicker—and watched what happened. Most of the sea surface got warmer, not surprisingly. But the cold tongue in the eastern equatorial Pacific actually got colder and longer. The model had been visited by La Niña.

And so, it seems, was the real world in the twentieth century. El Niño and La Niña alternated throughout the century, as they have for many millennia, but in an irregular way, with first one state being dominant for a few decades and then the other. For the last quarter of a century before the most recent drought, El Niño had the upper hand, which is one reason we started hearing about it so much. But when Cane, Clement, and their colleagues compiled sea-surface temperature data for the whole twentieth century, they detected a gradual long-term trend toward cool Niña-like conditions in the eastern equatorial Pacific—even as the ocean as a whole, and the planet as a whole, was warming.

The result is not so paradoxical, the Lamont researchers argued, if you consider the feedback loop that powers the ENSO cycle. When you add heat to the whole sea surface from above, the warm pool in the western Pacific warms up faster than the east, because in the east, the cold upwelling water spreads away from the equator and carries some of the added heat with it. Warm air rising off the now warmer pool in the west intensifies the atmospheric low-pressure zone there, strengthening the east-west pressure gradient that drives the trade winds. The stronger winds push more warm water to the west, completing the feedback loop. And thus does global warming give rise to the cold tongue of La Niña—at least in the Lamont model. This simple model considers the tropical Pacific in isolation. More complicated global models of the atmosphere and ocean don't all show the same effect. On the other hand, in covering the whole planet, those models may be missing some of the fine details of the tropical Pacific.

If global warming does favor La Niña, it will mean the tropical Pacific acts as a thermostat, counteracting the warming to some small extent. But it will also almost certainly mean an increase in drought in the West and the Great Plains. A team led by Richard Seager at Lamont has shown that the six major Western droughts in recorded history, including the Dust Bowl of the 1930s and the severe Southwestern drought of the 1950s, were all caused by La Niña. Seager and his colleagues were able to reconstruct global maps of sea-surface temperatures as far back as 1856, when the world's navies started measuring temperatures routinely. They then simulated the atmosphere's response to those temperatures, in the same way Hoerling and Kumar had done for the recent drought.

The Lamont workers found that tropical Pacific temperatures alone were enough to generate all the recorded droughts in the American West—including the one in the 1890s that brought an abrupt end to the settlers' myth that "rain follows the plow," and that led to the first federal irrigation projects in the West, such as the one that diverted the Truckee River and depleted Pyramid Lake. What's more, the temperature changes in the Pacific did not need to be large to create damaging decadelong droughts that had a large impact on American society. A few tenths of a degree Celsius were enough. The ocean did not have to be "perfect," as it was in 2002; it just had to be persistent.

Was it persistent enough in the Middle Ages to cause the megadroughts? Ed Cook's tree-ring data document the timing and extent of the droughts, but tree rings can't reveal how cool the water in the tropical Pacific was. Corals can, however. Corals have annual growth rings too, and the ratio of oxygen isotopes they incorporate into those rings depends on the temperature of the water—in cooler water they take up more of the heavy isotope, oxygen-18. By collecting fossil corals on Palmyra Atoll, in the Line Islands, and splicing their records together much as the records of individual trees are spliced together to create a long chronology, Kim Cobb of Georgia Tech has constructed a temperature history of the central tropical Pacific that goes back to AD 900 (though it still contains large gaps). Cobb's data suggest that the central Pacific was indeed relatively cool from the tenth into the thirteenth century, the period of the megadroughts. Conversely, during the Little Ice Age of the fourteenth to nineteenth centuries, when the glaciers were advancing in the Sierra Nevada and Mono Lake was at high levels, the corals indicate warmer, Niño-like conditions. Both observations fit the hypothesis that the megadroughts were caused by a persistent Niña-like state in the tropical Pacific.

What cooled the Pacific? The megadroughts happened during a time known by some as the Medieval Warm Period. Global-warming skeptics love the Medieval Warm Period, because its causes, whatever they were, were natural. In their view, that somehow disproves the notion that humans are warming the planet today—the current warming must be just one more peak of a natural cycle, and thus nothing exceptional. As the evidence has mounted that we are in fact warming the planet, the skeptics' fallback position has been that the Medieval Warm Period shows the consequences will be benign. Medieval warmth, they remind us, brought bountiful harvests and Gothic cathedrals to Europe and allowed the Vikings to sail the ice-free North Atlantic to colonize Iceland, then Greenland. A few centuries later came the Little Ice Age, and the Greenland colonists were eating dogs and ptarmigan feet until the last of them starved—but the Medieval Warm Period was a long stretch of salad days, or so the argument goes.

Some climate scientists, including the creators of the famous "hockey stick" graph that traces global temperatures over the past millennium, have countered the skeptics' argument by diminishing the Medieval Warm Period. They say it was merely a regional event confined to Europe and the North Atlantic. The hockey stick, which became the subject of its own distracting minicontroversy after the IPCC featured it so prominently in its 2001 report, shows no Medieval Warm Period. The shaft of the stick is a long, slow temperature decline into the Little Ice Age—which everyone agrees was global—and the blade is the sharp rise beginning in the late nineteenth century, as glaciers worldwide began to recede again. That trajectory definitely makes the current warming look exceptional—but perhaps just a bit too exceptional. Why did temperatures suddenly turn sharply upward in the late nineteenth century? Why did they continue to rise sharply in the first half of the twentieth, before we had put enough CO_2 in the atmosphere to make much of a difference to climate?

The Medieval Warm Period and the Little Ice Age have long looked to Broecker like different phases of the same natural cycle, and the most plausible cause of both at the moment is variations in the output of the sun—to that extent the skeptics may be right. But the skeptics are wrong to find that reassuring. While the Vikings were colonizing Iceland and Greenland, first the Mayan and then the Anasazi civilizations were collapsing in western North America. There is evidence that drought contributed to both events. Conversely, while the Greenland colony was shivering and finally perishing in the Little Ice Age, the West was enjoying a wet climate it would love to see again now. Climate change has always produced winners and losers, and no doubt it will in the future.

But the most important and unreassuring lesson to be drawn from natural climate cycles is that Earth's climate is extremely sensitive: it is capable of taking inputs that seem small to us and transforming them into outputs that seem large. The fluctuations in solar radiation that presumably helped precipitate the Medieval Warm Period and the Little Ice Age were no more than half a watt of energy per square meter of Earth's surface—the equivalent of turning on or off a single hundred-watt bulb in the average American house. No one understands how they could have had such large impacts on climate, but the evidence suggests they did. No one fully understands how fluctuations of a few tenths of a degree in Pacific sea-surface temperatures could cause long-lasting drought in the American West, but that's what seems to have happened. And the simplest hypothesis is that medieval warmth and medieval megadroughts were no mere coincidence—that the former caused the latter by promoting La Niña conditions in the Pacific.

Earth is already almost certainly warmer now than it was during the Middle Ages. Our CO_2 emissions, by trapping heat radiation from Earth, are already adding around five times more energy to the atmosphere than solar fluctuations did during the Medieval Warm Period. Maybe the current warming, instead of bringing a megadrought to the West, will be strong enough to overwhelm the feedback mechanism that turns warming into cooling in the eastern Pacific; the result might then be a more Niño-like future and more rain for the West rather than less. Maybe the American West will get lucky, in other words, and the worst effects of global warming will be felt elsewhere. But at the moment it does not seem likely.

As we write this, in the spring of 2007, the drought that began in 1999, and was interrupted in 2005 by a year of near normal rainfall in the West, seems to have returned. In March 2007, the runoff from the Colorado and other rivers into Lake Powell was a little above half of normal. In the first week of April, *The New York Times* reported on its front page on the various water projects—and water conflicts—that were taking shape as the West faced the prospect of a more arid future with a growing population. Utah was planning a $500 million, 120-mile pipeline from Lake Powell to the boomtown of St. George. Nevada was opposing that pipeline, but was planning one of its own to deliver water to Las Vegas from northern Nevada and Utah—which was of course fighting *that* project. In California on April 1, the Sierra Nevada snowpack was at only 40 percent of the long-term average, its lowest level since 1988, and state hydrologists were forecasting that the total runoff for the year would be only half of average. Governor Arnold Schwarzenegger was promoting a $4.5 billion bond issue to build more reservoirs. All seven Colorado Basin states—Wyoming, Colorado, Utah, Nevada, New Mexico, Arizona, and California—were negotiating with the federal government about what to do once the day comes when there is no longer enough water to be divided up in the way foreseen by the 1922 compact. Meanwhile the eight states bordering the Great Lakes were attempting to negotiate a regional compact of their own; its main purpose was to protect their vast store of freshwater from any attempts to grab it, for instance by the booming Southwest.

The same week as the *Times* article appeared, and the IPCC delivered its latest report predicting a world with more drought in it, yet another alarm-bell paper appeared in *Science*. The lead author was Richard Seager. The study, which was too late to be considered in the IPCC report, compared the forecasts for the Southwest made by nineteen different climate models. The forecasts were all based on one of the IPCC's middle-of-the-road carbon-dioxide-emissions scenarios, called A1B, in which emissions increase until 2050 as the world's population increases, then decline a bit as population declines and new technology is introduced, putting the CO_2 concentration in 2100 at 720 parts per million—a little more than two and a half times the preindustrial level. With that as input, only one of the models put out a wetter Southwest; the other eighteen showed the region getting dryer to various degrees. When Seager and his colleagues averaged all the forecasts together, they got a frightening ensemble. It indicated that a pronounced long-term drying may already have begun, and that by 2050, the normal climate of the Southwest will be as dry as the Dust Bowl of the 1930s.

The cause of the drying, the researchers found, was not a "perfect ocean for drought"—some of the eighteen models generated La Niña conditions in the tropical Pacific, and some didn't. Yet all eighteen predicted long-term drought. The drying, Seager's team concluded, was caused instead by the two other phenomena we mentioned earlier, both of them likely results of global warming: the tendency of the dry regions to get drier and the poleward expansion of the subtropics. To Seager and his colleagues, this result didn't negate the well-documented idea that La Niña causes Southwestern drought. On the contrary, it suggested a confluence of mechanisms, all pointing toward a drier Southwest:

> The drying of subtropical land areas that, according to the models, is imminent or already underway, is unlike any climate state we have

> seen in the instrumental record . . . The most severe future droughts will still occur during persistent La Niña events but they will be worse than any since the Medieval period because the La Niña conditions will be perturbing a base state that is drier than any experienced recently.

On the other side of the planet in early 2007, Australians were wondering whether they too were already seeing the future. In Australia, which lies in the subtropics of the southern hemisphere, it is El Niño that causes drought—but the drought that had beset the country since 2002 was more extreme than any other ever seen. Michael Coughlan, head of the Australian Bureau of Meteorology, attributed it to a "combination of short El Niño drought and longer-term decreasing rainfall." Farmers in the Murray-Darling basin, the southeastern basin that accounts for 40 percent of the country's agriculture, were facing the prospect of having their irrigation water cut off, because the Murray River and the Darling River were nearly dry. Prime Minister John Howard, a noted global-warming skeptic, called the situation "unprecedentedly dangerous." He said the drought had already cut 1 percent off the country's economic growth and suggested his fellow citizens pray for rain.

By the time this book appears, the Australian drought may have lifted; El Niño ended in 2007. But the long-term trend for Australia and for the planet seems clear. One reason the IPCC feels "high confidence" in its forecast of more drought worldwide is that it seems to be happening already. In 2004 Aiguo Dai and his colleagues at the National Center for Atmospheric Research calculated the Palmer index since 1870—from temperature and precipitation measurements, not tree rings—on a grid covering the globe. They found that since the 1970s, the area of the planet suffering from "severe" dryness had increased from 12 percent to 30 percent—at first as a direct result of the Niño cycle, but after that, the researchers concluded, simply because the warming atmosphere has been drawing more moisture out of the land.

Within limits, we can adapt to having less water. We can build more reservoirs, and we can conserve water, as Los Angeles has shown. Under pressure from drought, but also from court orders requiring it to return water to Mono and Owens lakes, the city has already cut per capita consumption from 180 gallons a day in the 1980s to 155 gallons a day in 2005—but that has merely allowed its total consumption to remain constant while its population grew by 750,000 people. Neither conservation nor new reservoirs can stop population growth, nor can they create rain or snow where none exist. It is not clear at all that the West as we know it could hope to adapt to a megadrought of the scale that occurred in the Middle Ages, especially if it were to start suddenly. It seems quite certain that other parts of the world—Africa, for instance—would not have the resources to adapt to such an event.

There is no proof that global warming will cause a megadrought, or a sudden sea-level rise for that matter. There is only a reasonable argument based on common sense—and on a metaphor. We have learned that Earth's climate has been capable of megadroughts and other extreme and abrupt fluctuations in the past, when given only a small push by the sun or by the Milanković cycles. It seems prudent to avoid giving climate a big push. If you're in a tippy canoe, you shouldn't dance—that's Richard Alley's version of the metaphor. If you're living with an angry beast, you shouldn't poke it with a sharp stick—that's Broecker's own favorite.

To avoid a dangerous reaction from the climate beast, we need in the next several decades to find a way to stop dead the increase in atmospheric CO_2, even as Earth's population and demand for energy are growing rapidly. We think there is a way.

THIRTEEN

Green Is Not Enough

By the time you reach Fort McMurray, a small but booming island in the great north woods of Alberta, you already feel near the end of the world. But to get to the Shell oil-sands mine you have to drive another forty-five miles north along the Athabasca River, through the lonely stunted forest of birch and poplar, spruce and jack pine. The oil sands—some people still call them tar sands, and tarry bitumen is what the sand is really laced with—lie under fifty thousand square miles of that forest in a layer that is two hundred feet thick. In most places the tarry layer is buried under another thousand feet of more recent sedimentary rock, and to get the tar out you have to drill wells, as into an ordinary oil reservoir, and pump steam down the wells to loosen up the tar. In the Athabasca Valley itself, however, the river has eroded away the overburden of sediments—with some help, at the end of the last ice age, from the Lake Agassiz flood. There the tar is so close to the surface, only ten or fifteen feet down, that it's strip-mined.

It's a mining operation like no other. When Shell takes visitors into one of the open pits at its Muskeg Mine, the large SUV is equipped with a twelve-foot lighted "buggy whip" to make sure it is always plainly visible and thus less likely to be accidentally squashed—and even so, the driver keeps a respectful distance from the giants in the pit, as if this were Jurassic Park and they were brontosaurus. In fact they are dump trucks, the largest trucks in the world, each one more than thirty feet wide and fifty feet high—trucks the size of small apartment buildings. If you were standing near one of the tires when it blew out, it would kill you. The trucks are dwarfed, though, by the cable shovel that works the face of the pit and fills a waiting truck in three or four scoops with four hundred tons of rock and mud. The loaded truck promptly turns and roars out of the pit, tires laboring at first for traction on the spongy floor. But before it reaches the conveyor belt, where it dumps its load in two seconds, it can get up to forty-five miles per hour, burning fifty gallons per hour of gas.

The conveyor belt never stops, even at midnight on Christmas Eve. It delivers fourteen thousand tons of ore an hour into a crusher, from where the rock travels to rotary drums that break it down still further, then into a giant separator where hot water parts the bitumen from the sand, causing the bitumen to float and the sand to sink. Later, solvent is added to make the bitumen more fluid, and it is pumped through a pipeline 250 miles south to Fort Saskatchewan, where Shell has an oil refinery. There, in a 140-foot-tall reactor with foot-thick-walls—so large it had to be brought from Japan via the Panama Canal and the St. Lawrence River, because it could not be hauled over the Rockies—the bitumen is injected with hydrogen at 435 degrees Celsius and two thousand pounds per square inch of pressure. The long carbon chains are cracked and converted into short-chain hydrocarbons; that is, a substance that had been suitable for use as roof sealer, and that the Native Americans had long used to waterproof their canoes, gets upgraded to a nice, light crude oil. Shell produces 155,000 barrels of oil a day this way.

All together, the oil-sands mines at Athabasca and elsewhere in Alberta produce more than a million barrels a day, and they are rapidly expanding. Other oil companies are getting into the business. India and China are investing billions. The cost of making oil from tar sand has fallen in the past three decades to around $20 a barrel, and in early 2008, oil was trading for around $90 a barrel. Tar sands are very good business now. The Alberta Department of Energy estimates that by 2015 its fields will be producing 3 million barrels a day. Alberta is sitting on oil reserves that it claims are second only to Saudi Arabia's.

When we hear talk of the end of oil, we think of Athabasca, and of those giant trucks roaring around in the Arctic winter night. Or we think of Appalachia, where American coal companies have since the 1970s been blasting away entire mountaintops and dumping them in the neighboring valleys, the better to expose the seams of coal there, still rich after more than a century of exploitation. Or we think of the oil companies that have moved beyond the continental shelf and are drilling in up to ten thousand feet of water, because it pays to do so; or of how those same oil companies are eagerly eyeing the Arctic, where sediment cores have suggested the possibility of abundant oil, which, thanks to the melting of the ice cap by global warming, may soon become accessible. The End of Easy Oil may be upon us now, and the End of Oil itself within a few decades. But the End of Fossil Fuels is centuries away. One reasonable estimate puts the total amount of fossil fuels at five thousand gigatons, most of it coal. There is enough to last for centuries.

That's true even though we are not slowing down our burning of it; we are burning it ever faster. In 1997, the year the world's industrial nations agreed at Kyoto to reduce their CO_2 emissions, global production of coal was 4.66 gigatons, or billions of metric tons. By 2004, it had risen more than 18 percent to a little over 5.5 gigatons. Two-thirds of that increase occurred in China, which has more than tripled its production since 1980 and now produces more than a third of the world's coal. China is fueling its breathtakingly rapid industrialization with coal. It is opening a new coal-fired power plant every week. That factoid has become such a staple of climate-change discussions that it has almost lost its power to astonish.

The effects on carbon dioxide emissions have been predictable. They are not decreasing, as the Kyoto protocol envisioned; they are increasing faster than ever. In the spring of 2007, an international team of researchers reported in the *Proceedings of the National Academy of Sciences* that global emissions from fossil-fuel burning had reached 7.9 billion metric tons of carbon, or 29 billion tons of CO_2. More alarming still, the emissions growth rate had accelerated dramatically, from 1.1 percent a year in the 1990s to more than 3 percent a year between 2000 and 2004. Most of the growth came in developing countries, which now account for 41 percent of the world's emissions. That is one of the hidden costs of globalization, and of all those cheap made-in-China products that clog the aisles of Wal-Mart and other Western superstores. Chinese manufacturing is much less energy-efficient than manufacturing in the industrialized countries, and the energy comes primarily from coal, which is the most carbon-rich fuel. All the CO_2-emissions scenarios that the IPCC devised as bases for its climate forecasts, even the ones that posit rapid growth in emissions, assume that the world economy will get steadily more energy-efficient and world energy generation will get steadily more carbon-efficient—emitting less carbon as clean-energy sources are adopted. That has indeed been happening in the United States and other developed countries. But it has not been happening lately in the world as a whole. Global CO_2 emissions are now growing faster even than the IPCC's worst-case scenario. We are headed off the charts.

We cannot blame the Chinese. In 2006, China passed the United States to become the largest CO_2 emitter—but per capita, its emissions remain less than a quarter of the American ones. Moreover, less than a quarter of all the CO_2 that has been put into the atmosphere since the nineteenth century was put there by China or by developing countries. It will take decades for them to catch up to our cumulative emissions—and since they still have a long way to go to catch up to our standard of living, they have every incentive to keep using the cheapest energy they can find. What that means is, no matter what happens in the West, the world's energy consumption is going to increase dramatically in this century, not decrease. And most of that energy will probably continue to come from fossil fuels, above all coal. They are cheap, readily available, and incredibly convenient to use—and we have a global infrastructure of power plants and refineries and pipelines and gas stations that is built around using them.

The End of Oil, when it comes, will not change that situation. We already know how to convert one fossil fuel into another—coal into gasoline, for instance. Germany did it when World War II cut it off from its oil supply. South Africa did it in response to the anti-apartheid trade embargo and today still makes 195,000 barrels a day of gasoline from domestic coal. The United States had a brief love affair with "synfuels" after the 1973 oil embargo, adopting a government program to subsidize coal-to-gas conversion. The economic rationale evaporated once OPEC lifted its oil embargo and the price of oil came back down, and the synfuels program is now widely viewed as a textbook example of wrongheaded government intervention in the free market—or at least it was viewed that way until recently.

Now the price of oil is up again and staying there. In the spring of 2007, legislators hailing from both political parties, but above all from coal-mining states, were pushing bills in the U.S. Senate that would create $10 billion in guaranteed loans to promote the construction of synfuel plants. The bills would also create a captive market for the new industry: the air force would be committed to buying jet fuel from synfuels. At the same time, with the Bush administration and its tenacious indifference to global warming drawing to a close, Congress was considering various bills that would for the first time attempt to restrain U.S. emissions of CO_2. Subsidies for coal, yet restraints on CO_2 emissions—there is a contradiction there.

Yet any realistic solution to the climate problem will have to resolve that conflict between the powerful drive to use fossil fuels and the real threat CO_2 poses. One way to get an idea of both the scale of the problem and of what an equitable solution might look like is to think in terms of a "carbon pie." The pie represents the amount of CO_2 we could still put into the atmosphere without disastrous effect. Its size is not easy to specify. We don't really know at what level the CO_2 concentration will become truly dangerous—at what threshold the climate might shift so that rapid melting of the ice sheets becomes unavoidable, say, or the intensity of the drought in the American West is incompatible with the civilization we have built there. James Hansen, director of the NASA Goddard Institute for Space Studies in New York, puts the threshold at 450 parts per million. The Goddard climate model predicts a one-degree-Celsius warming from that concentration, and Hansen thinks a global average temperature one degree warmer than today is enough to threaten the long-term stability of the ice sheets.

Hansen's views have great credibility in these matters. Testifying before Congress in 1988, he was one of the first researchers to say publicly that global warming was already happening, and he was right. He may well be right again that 450 ppm is the limit of safety for atmospheric CO_2. The drawback to setting that as a goal, however, is that it is probably not attainable. Right now every four gigatons of carbon we put into the atmosphere adds about 1 ppm to the CO_2 concentration. The size of the carbon pie that takes us to 450 ppm, 70 ppm above the current level, is thus only 280 gigatons. With global emissions from fossil-fuel burning now at nearly 8 gigatons a year, and with another 1.5 gigatons being added by deforestation, we will reach the 450 ppm limit in thirty years even if emissions don't accelerate—and right now they are accelerating.

A more realistic goal would be 560 ppm—a doubling of preindustrial CO_2—for which the middle-of-the-range climate-model forecast is a warming of three degrees Celsius. That would give us a carbon pie of 720 gigatons. How should the pie be sliced? The most equitable way would be for each country to get a slice proportional in size to its population. The industrialized countries as a group would then get around 20 percent of the pie, or 144 gigatons. At present they are emitting nearly 5 gigatons a year; at that rate, they will have eaten their pie in less than thirty years. Three decades to reduce their CO_2 emissions to zero: that gives an idea of the challenge those countries face, if they want to take full responsibility for the consequences of their prosperity and do as much as possible—though much less than some researchers advocate—to protect the planet from dangerous climate change.

Clearly, the industrialized countries are not going to reduce their carbon emissions to zero in thirty years. Most of them are not even going to meet their much less challenging obligations under the Kyoto Protocol, which calls on them to reduce their emissions by 2012 to below the 1990 levels—on average 5 percent below. The United States, which signed but never ratified the 1997 protocol, has not even tried to reduce its CO_2 emissions (although U.S. emissions did actually decline in 2006, thanks in part to a mild winter). At the same time, one of the great shortcomings of the Kyoto Protocol, which conservative climate skeptics have stressed and which has become starkly evident in recent years, is that it placed no obligations on developing countries.

The carbon pie suggests a conceptual way out of this dilemma. It dramatizes the reality that any solution to the climate problem is going to require an overarching deal between industrialized and developing countries. In essence, the former will have to buy extra pieces of pie from the latter, to avoid the choice between protecting climate and torpedoing their economies. In return, the developing countries will get some kind of help with developing—ideally, in a way that helps alleviate rather than aggravate the CO_2 problem. The bigger slices of pie that an equitable division would allot them would also allow them to use more fossil fuels for longer—which will in itself be an essential component of their development.

To be sure, the carbon pie simplifies reality. To stabilize atmospheric CO_2 at 560 ppm, the world will not really have to reduce CO_2 emissions all the way to zero in this century, because the ocean will continue soaking up CO_2, albeit at a steadily decreasing rate, for a couple of centuries to come. A more accurate way to state the challenge would be to say that by the second half of this century, the world will need to be putting much less CO_2 into the atmosphere than it is now, while the demand for energy will have doubled or even tripled. In other words, we will need to find a way to produce two or three times as much energy while emitting steadily less carbon than today—and ultimately, no carbon at all.

In the face of such a challenge, every little bit helps. In 2004, Stephen Pacala and Robert Socolow of Princeton University published an influential paper in *Science* that adopted that familiar maxim as a strategy. To stabilize atmospheric CO_2, they said—their target was 500 ppm—CO_2 emissions had to be frozen at their 2004 level of seven gigatons. But under a business-as-usual scenario, emissions would grow to fourteen gigatons over the next half century. Pacala and Socolow identified a long list of existing technologies, each one of which, they thought, was capable of being scaled up to a level of saving one gigaton of carbon emissions by the year

2054. Doubling the fuel efficiency of the world's cars from thirty miles per gallon to sixty miles per gallon, even as the number of cars is quadrupling to 2 billion, would constitute one such "stabilization wedge." Convincing people to travel five thousand miles a year in their cars instead of ten thousand would be another wedge. Doubling the number of nuclear power plants or increasing wind-farm capacity by a factor of fifty would be two more; but for photovoltaic electricity to save a gigaton of carbon, its capacity would have to grow by a factor of seven hundred. Pacala and Socolow identified fifteen stabilization wedges in all. To "solve the carbon and climate problem for the next half-century," they concluded, we just had to pick seven of them.

The limitation of this approach, as the Princeton researchers acknowledged themselves, is that we don't merely need to solve the climate problem for the next half century; we need to solve it for good. After 2054, energy demand will continue to grow, yet carbon emissions must continue to shrink. The gap between the emissions implied by the world's energy needs and the emissions that can be tolerated by climate will keep growing, and to fill it we will need more or bigger wedges. Yet most of the wedges identified by Pacala and Socolow are intrinsically limited; you can't make cars or buildings endlessly more efficient, and even if you could, it would not solve that large part of the emissions problem that has nothing to do with cars and buildings. Every little bit helps—but a strategy cobbled together of lots of little bits is not a reassuring response to the looming conflict between energy consumption and climate protection. We need a safety net that seems safer than that, one that seems capable of solving the whole problem.

None of the familiar sources of "clean" energy seem likely to be up to the task, at least not soon enough. Some will never be. There are not enough arable lands for biofuels, not enough volcanic hot spots for geothermal energy, and not enough dammable rivers for hydroelectricity ever to become more than small pieces of the solution to the climate problem. Only three technologies exist that have the potential to replace a substantial fraction of fossil fuels. The first is nuclear power. It's a measure of how urgent the CO_2 problem has become that some antinuclear environmentalists have lately been willing to reconsider their long-standing opposition. Nuclear power, as its proponents frequently remind us, has killed far fewer people than coal mining, not to mention pollution from coal-fired power plants. But that doesn't mean there aren't rational reasons to be skeptical of it. There is still no permanent disposal site anywhere on Earth for waste that will remain dangerously radioactive for thousands of years. And the fear that a proliferation of civilian power plants could promote the proliferation of nuclear weapons has only grown more acute since September 11, 2001. The huge expansion of nuclear power that would be necessary for it to contribute significantly to resolving the CO_2 problem is, in our opinion, just not going to happen.

Even wind turbines, those gentle green giants, are starting to provoke a backlash. The cost of wind power has dropped dramatically in recent years, to less than eight cents a kilowatt-hour—not as cheap as coal, but getting close. Wind-power capacity has been growing at around 30 percent a year. In the United States it still produces only around half of 1 percent of all electricity—about the same as the global total—but in Germany, which has subsidized wind power heavily, it produces more than 4 percent. In both countries, and especially in Great Britain, where people tend to be fiercely protective of their countryside, opposition to wind power is growing. Whether you think wind farms are pretty is subjective. But one of the things that incenses some opponents is the feeling that they're being asked to accept large blots on the landscape and are not getting a significant solution to the climate problem in return.

That argument has merit. Even if the biggest drawback of wind energy could be eliminated—being intermittent, it must always be backed up by more reliable fossil-fuel or nuclear plants—there is a fundamental limit to its exploitation that is not often discussed. Compared with what our demand for energy is likely to be in coming decades, there is just not that much energy in wind. If we were to satisfy a substantial fraction of that demand with wind power, we would be slowing the winds substantially, and that would slow the transport of heat and moisture around the planet. In other words, we would be replacing one climate impact with another that we understand even less well.

That leaves solar power, which probably is the ultimate solution to the climate problem. (If the tens of billions of dollars spent on researching nuclear fusion, the power source of the sun, ever produce a functioning reactor, then it too will become important—but not in the first half of this century.) Compared with wind energy, solar energy is abundant: the amount of energy the sun delivers to Earth's surface in an hour, around three hundred watts per square meter, is more than humans consume in a year. If we were to cover 10 percent of the Sahara desert with solar panels, we could produce around twice as much energy as the world consumes today. The trouble is, it would cost far too much. Producing electricity with photovoltaic panels costs around thirty cents a kilowatt hour today, and it costs a similar amount more to store the energy for the times when the sun isn't shining. That makes it around twenty times as expensive as producing electricity from coal. A generous program of tax rebates, such as California has recently instituted, can make it worth people's while to install solar panels on their roofs. Without such subsidies, solar power is nowhere near being competitive. Though the industry has been growing rapidly, and prices have dropped significantly, it still produces only around a hundredth of a percent of the world's energy. We could be wrong, but we doubt solar power is going to get cheap enough fast enough to keep us from more than doubling atmospheric CO_2.

Finally, the transportation sector accounts for nearly a quarter of global CO_2 emissions—it is the fastest-growing source—and no one has yet invented a practical nuclear-powered, wind-powered, or solar-powered car, let alone an airplane. The green solution to this problem is to use hydrogen as fuel. Whether you burn hydrogen in an engine or oxidize it in a fuel cell to make electricity, you produce only H_2O and no CO_2. But the hydrogen itself has to be made—and by far the cheapest way to make it is to steam coal, the process used in making synfuels. That process emits a lot of CO_2. Splitting water molecules with wind- or solar-generated electricity is the better way, but costs ten times as much.

Furthermore, there are at least two other fundamental reasons to doubt that the much touted hydrogen economy will arrive in time to protect us from climate disaster. First, it is proving extremely hard to design a tank that can hold enough hydrogen—

either as a liquid at –252 degrees Celsius, or as a gas at thousands of atmospheres of pressure—to allow someone to drive his car for three hundred miles without the nagging doubt that it might explode. The other obstacle is the infrastructure needed to deliver hydrogen to the consumer—the pipelines and filling stations. Because hydrogen is much trickier to handle, we would have to replace the enormous infrastructure now devoted to delivering gasoline—which grew gradually over a century—with a brand-new one. A study done at the Department of Energy's Argonne National Laboratory in Illinois estimated that it would cost $500 billion to build the facilities needed to supply hydrogen to just 40 percent of the cars in the United States.

Broecker had just begun thinking seriously about the climate problem during the 1970s when the first solar revolution hit the United States. In the wake of the 1973 oil shock, President Jimmy Carter created the Department of Energy and installed solar water heaters on the roof of the White House. They were dismantled during the Reagan administration. Solar energy has not gotten the government support it could have or should have gotten in the years since, but it has gotten *some* support, in the United States and in other countries. The fact that all the alternatives have been around for decades without managing to displace fossil fuels significantly is not only the fault of government or of oil and coal industry conspiracies. It is also the fault of the alternatives: they are just not as cheap and convenient as fossil fuels.

When you consider the gap between what the alternative energy sources seem capable of, and what needs to happen to keep the CO_2 concentration of the atmosphere from rising dangerously high, it is easy to get discouraged. Pessimism is a rational response. In the early 1980s Broecker taught an Earth-science survey course at Columbia, and after a few years, he decided to put his thoughts on the subject into a book. He called it *How to Build a Habitable Planet*. The book told the history of Earth, beginning with the Big Bang and ending with a chapter called "Mankind at the Helm." In that chapter Broecker talked about the CO_2 problem, which was increasingly on his mind. He wasn't sanguine about our prospects for escaping the consequences of our fossil-fuel habit:

> There is no practical way to capture the CO_2 produced when these fuels are burned . . . Nor can we reclaim CO_2 from the atmosphere. The amount of energy required to recover a kilogram of CO_2 would be comparable to the amount of energy gained by producing this CO_2 in the first place. Hence, to the extent to which we are dependent on energy, we are destined to have the CO_2 content of atmosphere rise . . . Since we can't significantly influence the amount of these gases that are produced or their fate once released, our strategy must be rather to prepare to cope with the changes the greenhouse effect is likely to bring.

Two decades later there is still no sign that we are capable of weaning ourselves from fossil fuels. Preparing to cope with the changes still seems wise—to the extent that we can. The changes seem more daunting now than they did in the mid-1980s, when the realization that Earth's climate had shifted abruptly in the past was only just beginning to dawn on Broecker and others.

Nevertheless, Broecker is more optimistic today than he was then. He no longer believes there is no practical way to capture CO_2 from the atmosphere. A colleague of his at Columbia, Klaus Lackner, showed him that isn't true. Thanks to Lackner, it now seems possible that, as far as fossil fuels are concerned, we really might have our cake and eat it too.

FOURTEEN

Scrubbing the Air

Broecker first met Klaus Lackner at a hotel in a suburb of Ottawa, in 1998 or 1999. They had both been invited by a Canadian think tank, along with other researchers from a wide range of disciplines, to talk about Canada's energy future. The conference as a whole did not make much of an impression on Broecker. Probably he gave a talk on the CO_2 problem and on the threat of abrupt change. Lackner got up and gave a short talk on how to solve the problem: he said we could just accelerate the natural process of geochemical weathering, in which atmospheric CO_2 reacts with magnesium- or calcium-rich minerals to make magnesium or calcium carbonates, which are inert and harmless. To do that we would have to mine billions of tons of such minerals and grind them to powder, the better to facilitate the reaction. Then we would have to find someplace to put the mountains of carbonate. Lackner called this CO_2-disposal process mineral sequestration. He made an immediate impression on Broecker. Broecker thought, This guy is nuts.

In fact, Lackner was a theoretical physicist, and a German one at that. More precisely, he was a product of the culture that gave the world twentieth-century theoretical physics, transplanted body and soul into the culture that gave the world twentieth-century industry. Lackner grew up in Heidelberg and studied theoretical physics at the university there. Feeling thwarted by the German seniority system, he got a postdoctoral fellowship at Caltech, where he worked with Murray Gell-Mann, the Nobel Prize–winning inventor of quarks. From there he moved to the Stanford Linear Accelerator Center, or SLAC—or "shlack," as Lackner still pronounces it in his perfect but slightly accented English. A couple of years into his American experience, he got a call from Germany telling him he had twenty-four hours to decide whether he wanted the faculty position that had been kept open for him. Lackner's mind was not troubled. He would have stayed in America, he says, even if he hadn't met a girl from Pasadena.

"One of the things that struck me," he says, "is that even as a brand-new postdoc, you are taken seriously here. It was far less hierarchical than in Germany. You treat people here differently based on what they can and cannot do. The price of being taken seriously is that you have to sink or swim—you can't jump in with a life preserver. That more than anything else decided it for me that I wanted to stay."

By the time Broecker encountered him, Lackner had moved to the Los Alamos National Laboratory, birthplace of the atomic bomb, and had moved up the ranks there to associate director. Leaving quarks behind, he had worked on supernovas, bombs, neural nets, and laser fusion. He thinks of himself as having moved away from theory and toward the real world of experiment—but his real-world work still had a certain big-picture theoretical bent. He gave talks on the future of technology. He wrote a conceptual paper about machines that build themselves. That was when he first thought about scrubbing excess carbon dioxide out of the atmosphere. He was looking for something that his concept machines might do.

The biggest problem confronting the world, Lackner was deciding at that time, is not whether quarks could exist in a free, unconfined state outside the atomic nucleus—the question that had exercised his brain as a theoretical physicist, and that he pursues these days as a hobby, the way other men might go bowling. The biggest problem was environmental. Malthus and his followers were right: we are headed for a brick wall. But the wall our growing population would soon crash into was not, as the Malthusians thought, the limited resources of the planet. It was the limited ability of the planet's thin biosphere to sustain the environmental impact our growing population and spreading industry are inflicting on it.

Ultimately that problem came down to energy. "If we had cheap, clean, and copious energy, we could solve our problems about being sustainable," Lackner says. Unlike other energy sources, fossil fuels are cheap and copious right now. If we could just make them clean, they would be perfect. The solution to that was technological; technology would allow us to become responsible stewards of the biosphere. And while these grand thoughts were coalescing in Lackner's mind, Broecker ran into him again in a more appropriate setting, at Biosphere 2, in the desert north of Tucson, Arizona.

Biosphere 2 was a stunt, a billionaire's folly, a visionary experiment, a magnificent research facility—it was all those things. But first of all it was a greenhouse. For a greenhouse it was immense, around three acres, but as a model of Spaceship Earth it was tiny. The idea its developers had sold to Ed Bass, the billionaire who put up the $150 million to build it, was that Biosphere 2 would reveal whether a closed system of that size could sustain human life for long periods. The answer turned out to be—not pleasantly. The eight Biospherians who were locked into the facility in September 1991, embarking on a sort of mission-to-Mars on Earth, started losing weight rather quickly. Eight months later, the director of the facility sought out Broecker, author of *How to Build a Habitable Planet*, for advice on an even graver problem: the oxygen level in the greenhouse had plummeted from 21 to 14 percent. It was as if the hungry Biospherians were living in the thin air at seventeen thousand feet.

Broecker realized immediately what the matter was. The designers of Biosphere 2, in their concern that the residents be able to feed themselves, had overloaded it with organic-rich soil. Bacteria in the soil were consuming oxygen and respiring CO_2. The plants in the greenhouse couldn't photosynthesize fast enough to take up all the CO_2, especially in winter; the concentration had thus risen to eight times the level in the outside atmosphere. It would have risen far higher still were it not for two things. Knowing that they would have at least a small problem with excess CO_2 in winter, the management had installed a scrubber in the greenhouse, which removed CO_2 from the air by reacting it with sodium hydroxide. And as Broecker's graduate student Jeff Severinghaus eventually discovered, even more of the CO_2 was being absorbed by the concrete walls at the base of the greenhouse. It was reacting with calcium hydroxide in the outer layers of the concrete to form a gradually deepening rind of calcium carbonate. Biosphere 2 had a small-scale mineral sequestration program going on.

The facility's original mission was a fiasco—the greenhouse residents nearly starved, and tank trucks of oxygen had to be delivered to rescue them from slow asphyxiation. Eventually Ed Bass fired the managers and, as an indirect result of Broecker's connection to the project, contracted with Columbia University to take over Biosphere 2 and use it for research and education. Broecker hoped that Biosphere 2 would become a laboratory for testing the impact of the ongoing rise of CO_2 in the original biosphere outside the glass walls. And some valuable research did get done along those lines before Columbia too bowed out of the project. It now stands forlorn in the desert, waiting once again to find its purpose.

While Columbia was managing the facility, however, Klaus Lackner joined the scientific advisory committee that Broecker served on. In that more relaxed context, Broecker found Lackner's tendency to think big more exciting than crazy. Lackner is a big, relaxed man, with a mostly bald head, rounded cheeks, and a disarmingly open smile. He has the theoretical physicist's penchant for attacking problems from a base of first principles, but he is patient with people who are less well-grounded in those principles—thermodynamics, say—than he is. The lack of arrogance in someone so obviously smart is charming. As Broecker talked with Lackner at length, his initial impression—this guy is nuts—mutated into something different: this guy is brilliant. Soon Broecker was trying to lure Lackner away from Los Alamos to a job on the Columbia faculty.

When academics interview for a new job, they give talks to show their stuff. At Columbia, Lackner chose to talk about capturing carbon dioxide out of the air. There was nothing revolutionary about the idea of capturing CO_2 per se; everybody knew it could be done. It was being done at Biosphere 2. It was being done on the space shuttle and in every submarine in the world to keep the crew from asphyxiating from their own exhalations. It was being done in a variety of industrial settings—at natural gas wells, for example, where CO_2 must be removed from the gas to increase its value as fuel and prevent corrosion in the pipeline. And by the late 1990s the idea that we might capture CO_2 from the flue gases of coal- or gas-fired power plants, to keep it from entering the atmosphere, had gained wide currency. What was novel about Lackner's idea was the scale—he was proposing to scrub CO_2 out of the atmosphere itself, not out of a small, enclosed space or a confined stream of gas—and also his way of conceptualizing the task to show that it was doable.

Carbon dioxide makes up anywhere from a tenth to a sixth of the flue gas at a power plant, but its concentration in ambient air is more like one in three thousand—that is, it is three hundred times more dilute. To a chemical engineer, that ends the discussion. It would be far too expensive and impractical, goes the conventional wisdom, to try to cleanse the atmosphere of such a dilute pollutant. Lackner looked at the CO_2 in the air differently: in terms of the energy content of the fuel that had been burned to generate it. If you extract a certain amount of CO_2 from the air, he reasoned, you could replace that same amount by burning a fossil fuel, without harming the planet. If the fuel is gasoline, then the gasoline needed to produce the amount of CO_2 in a cubic meter of air would deliver ten kilojoules of energy. But that same cubic meter of air, blowing at a brisk ten meters per second, or 22.5 miles per hour, contained only fifty-eight joules of kinetic energy. In other words, there was 170 times more energy to be gained from extracting the CO_2 from air than from extracting the wind.

Put another way, to satisfy the energy demand of the average American, you would need a windmill that swept out an area as large as the side of a barn. If instead you supplied that energy with fossil fuels, and extracted the resulting CO_2 from the wind, you would need an area the size of a barn window. Even if you were able to extract only half the CO_2, your apparatus would be nearly one hundred times smaller than the equivalent windmill. And yet wind energy is economically competitive today.

Scientists and engineers who hear this argument, but who have not really looked into the matter themselves, are often skeptical. Lackner's idea violates their intuition. You can't fight thermodynamics, they often say. "It's a great way to end a discussion," Lackner says. A few years ago the IPCC convened a special committee to prepare a massive report on all aspects of carbon capture and sequestration. The report focused exclusively on capturing CO_2 at power plants and other point sources. It discussed in one sentence the idea of capturing it from air: "The possibility of CO_2 capture from ambient air (Lackner, 2003) is not discussed in this chapter because the CO_2 concentration in ambient air is around 380 ppm, a factor of 100 or more lower than in flue gas." As far as the committee was concerned, that was enough said.

Lackner was on the committee himself. It rankled a bit to hear his idea dismissed with arguments from first principles, because he feels he knows the principles. He has calculated, from thermodynamic first principles, the energy required, in theory, to extract CO_2 from the atmosphere: it is around twenty kilojoules per mole. (A mole of CO_2 is forty-four grams, about an ounce and a half, and twenty kilojoules is about five calories, the energy contained in a stick of sugarless gum.) The energy required to extract CO_2 from a concentrated flue gas is about five kilojoules less—in theory. But in practice the systems that do it spend sixty to a hundred kilojoules, which is way more than the thermodynamic minimum for extracting it from either flue gas or the open air. "The thermodynamic argument is plain wrong," Lackner says. "It takes more energy to extract CO_2 from air than from flue gas, but the difference is quite small."

First principles are not the only ones, he decided. They weren't going to win this argument for him. The way to convince the skeptics was to actually build a device that could extract CO_2 from air economically. While he was listening to the arguments on the IPCC committee, Lackner was already getting a chance to do that, thanks in part to Broecker.

By 2003, Lackner was on the staff of Columbia, but Columbia was preparing to pull the plug on Biosphere 2—turning it into a satellite campus had not worked out financially. At Biosphere 2 there was an engineer whom Broecker had helped recruit named Allen Wright. Wright had an unusual background that made him a kind of polar opposite to Lackner. He grew up near Detroit, son of an automotive engineer at Ford, dropped out of Northern Michigan University after two years, then spent a decade working in construction. First he drove a truck. Later he managed and even designed the installation of electrical systems. Eventually, feeling he needed a college degree, he quit his job and enrolled at Michigan State, where he completed his initial major—which was in fisheries biology. When he graduated, the only job opening he saw in that discipline involved, as he recalls, "sitting on a rock in the Aleutian Islands to count seals." Wright never worked as a fisheries biologist.

Instead, because he happened to walk by the right dock on Grand Cayman on the right day, he became a submersible pilot and engineer. By the time he applied for a job at Biosphere 2—his parents and older brother had settled in Tucson—he had spent a decade working for the submersible program at the University of Hawaii, where he had dived into volcanic craters on the seafloor. He came recommended by the scientists there as a miracle man with a knack for making things work. At Biosphere 2 his job was to help scientists make their experiments work. One of the people he met there was Klaus Lackner, who kept bending his ear about testing an idea he had for capturing CO_2 out of air. It was the beginning of a beautiful friendship in which Lackner does most of the talking, and Wright most of the building.

By the fall of 2003, when Columbia withdrew from Biosphere 2, leaving Wright jobless, Broecker had become convinced that Lackner's air-capture idea represented the best hope for preventing dangerous climate change, and that the best hope for getting it off the ground was for Lackner and Wright to collaborate. Broecker had also become a friend and adviser to Gary Comer, who had already agreed to fund arcane research into past climate changes. A billionaire entrepreneur who could see the interest in sediment cores, Broecker decided, would probably be even more interested in helping two guys in a garage in Tucson save the world. Broecker suggested that Wright start a company to build an air-capture device, and that he and Lackner approach Comer for venture capital.

A few months later Wright found himself in a conference room at the airport in Teterboro, New Jersey, outside New York, waiting for Comer's jet to land. With him were Lackner and Broecker, who had informed him at breakfast that morning that he, Allen, would be doing most of the talking. With him too was his older brother, Burt, a former Tucson fireman. Burt had grown up taking apart cars in his father's backyard, still drives a '62 Chevy pickup, and now runs a small company that designs and installs sprinkler and ventilation systems. "I'm a mechanical guy," he says. "I build things and I get great satisfaction out of that." He had no previous interest in the environment and no idea what he was doing in the room, except that he was intimately familiar with all the mechanical components—the pipes and pumps and valves and nozzles—that a CO_2 air-capture device would inevitably include. Also, his mere presence was a help. As Broecker pointed out, it meant that the world-changing air-capture device would be built by the Wright brothers, Allen and Burt replacing Orville and Wilbur. How could Comer resist?

That morning in Teterboro, Comer arrived with his finance man, Bill Schleicher. Lackner ended up talking for a while, because after all the air-capture device was his idea. Allen talked about the cost of building the device. Burt talked about his company. Finally Comer turned to Schleicher for his opinion. "This is way beyond venture capital," Schleicher said. "This is *ad*venture capital." Comer said, "I don't see why it couldn't work." He and Schleicher left. Allen Wright turned to Lackner and said, "What just happened?" Lackner said, "Congratulations. You got funded."

After a whirlwind tour of Manhattan, Lackner took the Wright brothers back to his office at Columbia, where he started to talk at them. At around nine that evening Burt heard a noise: the sound of his brother's head hitting the table. "You have to stop," Allen said to Lackner. "There's no more room." The next day the Wrights boarded a plane back to Tucson. They had $5 million to build a machine that would scrub CO_2 out of the atmosphere, and three years in which to do it. "Probably for the first forty-five minutes we just stared at the back of the seat in front of us," Burt recalls. "And then Allen said, 'Oh my God, what have I done? How can we pull this off?' "

They started with sodium hydroxide, or lye, because they knew it would work. It is a strong base and nasty, caustic stuff, but for that very reason it wants to grab CO_2, which is acidic. If you expose liquid sodium hydroxide to air, it will react with CO_2 in the air to make sodium carbonate. The Biosphere's scrubber used that reaction. To get the sodium hydroxide back, you react it with calcium hydroxide, which exerts an even stronger pull on the carbon—the result is calcium carbonate, which is what limestone is made of. At the Biosphere, they just stored the calcium carbonate in drums.

That would not be practical on the scale Lackner was thinking of—the scale of the whole atmosphere, with hundreds of thousands of scrubbers spread around the planet. If you had to keep trucking in calcium hydroxide and trucking out calcium carbonate for each one of those scrubbers, the process would never be economical. Each of the scrubbers needed to be self-sufficient—it needed to deliver pure CO_2 out the bottom for disposal, and it needed to regenerate its sorbents in pure form to be recycled back into the scrubber. An easy way to strip CO_2 from calcium carbonate, one the cement industry uses to make cement, is to heat the calcium carbonate to nine hundred degrees Celsius. Lackner and the Wrights knew they couldn't afford that much energy. But they wanted to start with sorbents that worked, because they wanted to focus on the design of the scrubber itself.

The need to save energy was the paramount design consideration, because the operation of the scrubber had to be as cheap as possible, and also because any electricity it used might be generated with fossil fuels, which would just put more CO_2 into the atmosphere. To avoid canceling its own efforts, the scrubber designed by Global Research Technologies (GRT)—the company of which Allen Wright was now the president, with an office sublet from his brother, in between a beauty salon and an alternative-healing center in a Tucson strip mall—had to be much more energy-efficient than the scrubbers previously built for industry or for Biosphere 2. It could not use giant fans, for instance, to blow air over the sorbent; natural wind would have to do.

Unlike industrial scrubbers, such as the ones that remove CO_2 and other contaminants from natural gas, the GRT scrubber would not try to remove all the CO_2 from the air in a single pass. The important goal was not to get all the CO_2 out of a given batch of air—"There's lots of air out there," says Allen Wright—but instead to convert all of a given batch of sodium hydroxide to sodium carbonate. After the sorbent has collected CO_2, it has to be stripped of CO_2 and then pumped back into the collector. This recycling phase, Lackner had realized from the start, would account for most of the energy and operating cost. Any sorbent that was recycled without having reacted with CO_2 just wasted energy. The scrubber had to expose as much sorbent as possible as quickly as possible to the CO_2 in slow-moving air.

In February 2004, in a small warehouse outside Tucson, not far from the Tucson Electric coal-fired power plant, Allen Wright and a small team went to work. Periodically Lackner would breeze in for a few days and keep Wright up to all hours talking about design ideas. They tried a field of hanging plastic strips with sodium hydroxide dripping down them; they tried a stack of plastic wheels with sodium hydroxide fountaining from the hub and flowing down through the spokes; they tried venetian blinds from Target. "The pile of Smithsonian exhibits grew at an alarming rate," says Allen Wright. He is a modest, self-deprecating man, quick in conversation to downplay his own expertise. But Broecker's Wright brothers quip had stuck with him.

By the fall of 2004 the Wrights and Lackner had a contraption they were ready to show off to Comer and Broecker. It was simplicity itself: a line of hanging plastic sheets, eight feet tall by four feet deep and spaced an inch or so apart, such that the open ends of the device were about a foot wide. As air blew through the gaps between the sheets, it came into contact with the sodium hydroxide, which was sprayed onto the tops of the sheets through closely spaced nozzles, such that it cascaded down them in a uniform layer. The spacing of the sheets kept the device from presenting too much resistance to the air and kept the air flow smooth, minimizing the turbulent eddies that would have reduced the efficiency of the chemical reaction. At the bottom, a tray collected the liquid again, which was a mixture of sodium carbonate and water. The device worked: in converting sodium hydroxide to sodium carbonate, it extracted ten kilograms of CO_2 from the air every day.

But it was just a first crude proof of principle. And as Lackner and the Wrights would come to appreciate over the next two years of furious trial and error, collecting CO_2 was the easy part; separating it from the sorbent was harder. Sodium hydroxide was such a strong sorbent and bound the CO_2 so tenaciously as sodium carbonate that there seemed no way to strip the CO_2 off again without expending prohibitive amounts of energy. Sodium carbonate can itself react further with CO_2 to form sodium bicarbonate. But that reaction is too slow to make sodium carbonate useful on its own as a sorbent—essentially, it binds CO_2 too weakly.

Lackner and Allen Wright are cagey about the solution they eventually found—for the simple reason that, although they are trying to help save the world, they would like to make a profit too. In early 2006, speaking to an audience of Icelandic scientists—Iceland may be a good spot to store CO_2 underground—Lackner made the case for air capture in the most general terms, refraining from giving any details about the device GRT was working on. "It's not rocket science," said one member of the audience, pointing out that even a glass of water left out overnight will absorb some CO_2. Lackner agreed. "And that, by the way, is why the company wants me to keep quiet about it," he said. In the spring of 2007, when GRT announced in a press release that it had a practical air-capture device, it provided no details of how the device worked.

An invention can be ingenious even if it isn't rocket science and is hard to protect with patents. Allen Wright and Lackner have found a plastic, commonly used for other purposes, that binds CO_2—exactly how is not clear. In GRT's latest device, the hanging sheets are made of this kind of plastic, which has a mesh texture. Instead of being bathed in sodium hydroxide, the sheets are kept dry as air flows through them. Once the plastic has become saturated with CO_2, the doors of the device are closed, and the sheets are rinsed from above with sodium carbonate—not sodium hydroxide. The sodium carbonate reacts with the CO_2 to form sodium bicarbonate, which is baking soda. A brine of baking soda and water accumulates at the bottom of the collector and is channeled to the CO_2 separator.

The beauty of the scheme is twofold. First, baking soda and sodium carbonate are harmless, whereas sodium hydroxide will take your skin off. Second, it takes a lot less energy to separate CO_2 from baking soda and regenerate sodium carbonate than it does to go from sodium carbonate—or calcium carbonate—back to sodium hydroxide. Baking soda is a salt, and there is a well-developed technology for separating a salt back into its constituent ions, which have opposite electric charges. You put a series of membranes in the brine, with a positive electrode at one end and a negative electrode at the other. In the GRT separator, sodium carbonate accumulates on one side of the membranes and CO_2 on the other.

The device standing in GRT's warehouse in the spring of 2007 was what Allen Wright called a pre-prototype—actually two of them were standing side by side, each eight feet by four feet by one foot, like the original model. The proud inventors were now debating what the next step should be. For Lackner the scale of what was needed was clear: it needed to be something like Altamont Pass. The construction of that immense wind farm east of San Francisco in the 1970s, Lackner says, was the proof that wind energy was serious business. The air-capture version of Altamont Pass, he says, would be a farm of a hundred or so collectors, each removing one ton a day of CO_2 from the air—roughly the total production of two thousand Americans. To build such a project, GRT was looking to move beyond Comer's adventure capital to find some actual venture capital.

A one-ton-a-day CO_2 extractor, Wright says, could fit in a forty-foot shipping container. What exactly it would look like is still an open question. When Lackner first dreamed up the idea, he imagined a sort of football goalpost, swiveling in the wind, with venetian blinds spanning the uprights and sorbent cascading down the blinds from the top; he also imagined a big structure, capable of extracting 90,000 tons of CO_2 a year instead of 365. A Canadian documentary once showed the scrubber towering like an apartment building over Central Park. More recently, a GRT engineer has imagined the scrubbers as looking more like the Leaning Tower of Pisa—but straight and only twenty feet high—such that wind might pass through them from any direction. Lackner and Wright are continually looking at new geometries for their CO_2-binding plastic, new ways to rinse the CO_2 off it, and new ways to recover the solvent.

All the CO_2 scrubbers would not have to look the same, any more than all CO_2-producing cars look the same. But keeping them small enough to be mass-produced seems like a good idea, says Wright. No matter how big they are, the numbers of them needed to make a dent in global warming will be large—although again, not nearly as large as the number of windmills that would be needed to have an equivalent effect on atmospheric CO_2. The world now produces from burning fossil fuels nearly 29 billion tons a year, or 80 million tons a day, of CO_2. To pull all of it back out of the atmosphere would thus take around 80 million ton-a-day collectors.

There would be no reason to capture it all, however. It will usually make more sense, for instance, to capture the emissions from stationary sources, such as power plants, directly at the smokestack. Point-source carbon capture seems virtually certain to happen on a large scale; the U.S. government, convinced that coal is central to the country's energy future, has been heavily promoting the idea. In 2009, a consortium of companies organized and partially financed by the Department of Energy plans to begin building a new coal-fired power plant that will capture its CO_2 emissions. The project has been dubbed FutureGen, and as if that name weren't bad enough, it will use a technology called integrated gasification combined cycle. In an IGCC plant, coal powder burned only partially, at high pressure and in the presence of steam, creates a "syngas" that is mostly hydrogen and carbon monoxide. The gas is then burned in a turbine, and the hot exhaust from that turbine is used to make steam that drives a second turbine. The two turbines make the IGCC process more efficient than a conventional coal-fired plant.

The extra efficiency can help pay for carbon capture, which is easier to do at an IGCC—the carbon monoxide can be oxidized further to CO_2 and separated out of the syngas even before it is burned. Conventional coal-fired plants can be retrofitted with CO_2 scrubber towers, but they would end up having to burn so much extra coal to power the scrubber that it would raise the cost of electricity by as much as 60 percent. With an IGCC, that price rise would only be 30 percent or so, according to a recent MIT report on the subject. At least one forward-looking American utility, American Electric Power of Ohio, is already planning to build two IGCC plants, to be ready for the day when the government begins to regulate carbon emissions. And outside Berlin, Germany, the Swedish power company Vattenfall is building a thirty-megawatt pilot plant to test a different approach to capturing CO_2, by burning the coal in pure oxygen.

But nearly a quarter of CO_2 emissions worldwide come from mobile sources—cars, trucks, ships, planes—and their share is rapidly growing. (In the United States, where there are far more cars than drivers, a third of CO_2 emissions come from the transport sector.) Another 10 percent or so of emissions come from heating residential and commercial buildings. Thus more than a third of CO_2 emissions worldwide, and more than 40 percent in a country like the United States, cannot readily be captured at the source. Even if you could build a CO_2 scrubber the size of a catalytic converter, you could not attach it to an automobile tailpipe. The gasoline in a twelve-gallon tank weighs about one hundred pounds, but when you burn it and oxidize the carbon, it produces more than three hundred pounds of CO_2. There is not room to store that CO_2 in a car, nor in a plane for that matter. And if you did store it, where and how would you dump it? The only practical place is where it goes now—into the air.

We cannot solve the CO_2 problem without tackling small and mobile sources. Right now Lackner and Wright's invention offers the only hope. Capturing CO_2 out of the air has a huge advantage: the scrubbers can be put anywhere. When CO_2 is collected from the flue gases of power plants, pipelines will be needed to carry it, perhaps hundreds of miles, to wherever it is to be disposed of. With air capture, the atmosphere is the pipeline. Since CO_2 is quickly mixed through the whole atmosphere, taking it out at one place lowers the concentration everywhere. The scrubbers don't have to be near the sources of the CO_2. They can be built right above the disposal site.

FIFTEEN

Disposing of Carbon

Disposing of the CO_2, of course, is a problem in and of itself—a much harder problem, in Lackner's view, than capturing it in the first place. The quantities are stupefying. Before it is put anywhere, the CO_2 has to be compressed to a liquid, so that it takes up less space and is less likely to leak. If the twenty-nine gigatons produced by the world's fossil-fuel burning in a single year were liquefied and spread over Manhattan, they would bury the island to about the eighty-fifth floor of the Empire State Building. In a little over sixteen years the CO_2 would fill Lake Erie. The disposal problem is large.

The CO_2 cannot be put in a lake at the surface, because it would just evaporate back into the atmosphere. It has to be kept at high pressure. That means it has to be stored at great depth beneath Earth's surface, either in the ocean or on land. In some ways the deep ocean makes most sense, at least in the short term. The CO_2 we put into the atmosphere will eventually enter the deep ocean anyway, but the process is slow—basically because the oceanic conveyor belt is slow. It takes hundreds of years for CO_2 to mix completely through the deep. Right now it is accumulating in and slowly acidifying the surface waters, where most of the life in the ocean is, and it is reacting there with dissolved carbonate ions to make bicarbonate, much as it does in the Tucson scrubber. The process keeps atmospheric CO_2 lower than it would otherwise be, but it is bad news for all the organisms near the surface that make shells of carbonate—corals, for instance. In the indoor coral lagoon at Biosphere 2, Chris Langdon of Columbia showed that the reduction in dissolved carbonate caused by rising CO_2 slows the growth rate of corals significantly. In the real ocean, this effect adds to the stress imposed on corals by global warming itself. It is not yet known how rapidly corals can adapt to these changes; if they can't, and if we don't dramatically curtail the CO_2 buildup, the future of the world's reefs looks doubtful.

A better solution than watching their demise would be to capture atmospheric CO_2 and pipe it directly into the deep sea—putting it where it will eventually go, but without letting it do damage at the surface first. Liquid CO_2 is slightly less dense than water at the surface, but it is more compressible, and below a depth of three thousand meters, or ten thousand feet, the pressure is high enough that the CO_2 becomes denser than water. If it is injected below that level, it will sink to the seafloor. Peter Brewer of the Monterey Bay Aquarium Research Institute has done experiments showing what happens to it there. Using MBARI's robotic submersible, he injected up to one hundred pounds of liquid CO_2 into containers on the seafloor more than two miles down in Monterey Canyon. On Brewer's videotapes, you can see fish swimming up to the open containers and lingering there curiously. Brewer has also placed fish and other deep-sea animals in cages next to the containers. Nothing much seems to happen to them, but the experiments have been too limited in time and volume to say much about the ecological effects of a serious CO_2 injection. Some biologists are concerned that deep-sea animals, because of their slower metabolism, might be particularly vulnerable to the effects of acidification.

What happens to the liquid CO_2 itself is clear: it quickly reacts with the water to form a solid called a clathrate, with seven water molecules forming a cage around each molecule of CO_2. The clathrate is denser than either water or liquid CO_2; it forms a kind of slush on the seafloor. Over time the slush slowly dissolves, and the liberated CO_2 spreads through the deep. Were CO_2 ever to be injected into the deep on a significant scale, the worms and other organisms living directly under the CO_2 slush pile would surely be killed—but then, we slaughter many earthworms and other organisms under every road, runway, and parking lot we build on land. The bigger concern is the effect the CO_2 would have on the ecosystem as it dispersed through the abyss and was neutralized by reacting with carbonate. The pattern of the dispersal would depend on the bottom currents. Pilot studies would be needed to gauge the impact.

Such a pilot study has never been done; Brewer's little experiments are about all there is. Back in the 1990s researchers from MIT and Japan attempted to organize a more substantial but still small test injection of CO_2 from a pipeline off Hawaii; opposition from environmentalists killed the project. Some people seem to have a visceral reaction against the disposal of anything in the ocean. Perhaps because they are horrified at what we have done to the land, they hope to keep the deep pure and untouched.

To marine chemists such as Brewer, who participated with Broecker in GEOSECS as a young man and made some of the first direct measurements of the ocean's uptake of fossil-fuel CO_2, that reflexive opposition is frustrating—because it seems to ignore the essence of the CO_2 problem, which is that it leaves no part of the planet untouched. "The ocean is absorbing very large amounts of CO_2 right now," Brewer told the audience at a conference on carbon sequestration in January 2005. "We have put about five hundred billion tons of fossil fuel CO_2 in the ocean, and global disposal today is at the rate of about one million tons per hour. It's mankind's largest CO_2 disposal activity." Ocean sequestration could never be a panacea; for one thing, about 15 percent of the CO_2 would eventually escape back to the atmosphere, once the deep water came back to the surface in such places as the Antarctic and the North Pacific. But that, given the speed of the conveyor belt, would only happen after a few centuries, when presumably we would be through with fossil fuels, and atmospheric CO_2 would have started to come down again. As Brewer puts it, ocean sequestration would allow us to "shave the peak" off the atmospheric CO_2 curve.

He is not optimistic that it is going to happen. "Whenever we sit around over beers at one of these conferences," he said over coffee after his talk in January 2005, "we talk about where we think we'll be fifty years from now. To a man, the answer is, CO_2 will still be rising. Climate will have changed. We'll have gotten more efficient at energy generation, but our population will have risen. We'll have developed sequestration technologies—and maybe they'll be doing three percent of what's needed."

Maybe Brewer is right, but we hope not. Attitudes toward carbon sequestration have lately been evolving along with attitudes toward global warming itself. As industrialists and politicians who once denied the problem have been forced to admit its reality, environmentalists who were once opposed to anything but pure-green solutions have become more realistic as well, in the face of the urgency of the problem. It's too soon to say right now whether purposeful ocean sequestration will remain a complete nonstarter. But in any case, the ocean is not the only place to put captured CO_2. There is a lot of room underground as well.

Carbon dioxide is being disposed of underground already—on land, but also under the seafloor. Since 1996, the Norwegian oil company Statoil has been injecting a million metric tons of CO_2 a year under the floor of the North Sea at its Sleipner natural gas field. The gas contains about 10 percent CO_2, which has to be extracted to allow the methane to burn well. At other gas fields in the world, the CO_2 is just vented to the atmosphere. But Norway taxes CO_2 emissions at $50 a ton. To save $50 million a year in taxes, Statoil invested $80 million in equipment to capture the CO_2. On the offshore platform, the natural gas extracted from the seabed is pumped into two scrubber units—115-foot-tall towers filled with steel pellets. As the gas percolates up through the pellets, it encounters liquid monoethylamine percolating down. The MEA captures the CO_2, which is then released again (by heating) and compressed to eighty atmospheres. In that "supercritical" state, neither gas nor liquid but very fluid, it is injected back into a layer of sediment that lies above the natural gas deposit, half a mile below the seafloor.

That layer is called a saline aquifer—a layer of sedimentary rock, in this case sandstone, whose pores are filled with brine. At the top of the aquifer there is a cap of nonporous shale. The supercritical CO_2 is less dense than the brine, and after it is injected, it floats upward through the aquifer until it hits the cap. Seismic surveys that track its progress—sound waves sent down from the surface bounce off the CO_2 layer, because it is less dense—show that so far the CO_2 has not breached the cap. If it does, there are shallower seals to impede its upward progress. Over time it should dissolve in the brine, which would thereby become heavier and sink back to the bottom of the aquifer.

In the United States, 7 million tons a year of CO_2 is being injected underground right now—but not, of course, because there is a carbon tax. The CO_2 is being pumped into the ground to help get more carbon out, in the form of oil. The process is called enhanced oil recovery. After an oil company has pumped all the readily pumpable oil out of a reservoir, then flooded it with water to get even more, it typically has only extracted 30 to 40 percent of the total. Flooding the reservoir with CO_2 recovers another 10 or 15 percent; the CO_2 either pushes the oil toward the well or dissolves in it and makes it less viscous and hence more fluid. Most of the CO_2 stays underground, in a reservoir that has safely held oil and gas for tens of millions of years without substantial leakage.

Oil companies actually buy CO_2 for this purpose and have done so for the past few decades. At its Shute Creek Gas Plant in western Wyoming, Exxon-Mobil extracts nearly 5 million tons a year of CO_2 from a natural gas field that contains 50 percent of the unwanted gas—at least, it used to be unwanted, and Exxon-Mobil used to just vent all of it into the atmosphere. But for the past two decades it has pumped some of the CO_2 into a pipeline that runs hundreds of miles east to the Powder River Basin in Wyoming and also into northwestern Colorado. There it sells the gas, at a price rumored to be $10 a ton, to other oil companies that use it for enhanced oil recovery.

A similar project, this one under the auspices of the International Energy Agency, is under way in North Dakota and southern Saskatchewan. A two-hundred-mile pipeline transports CO_2 from the Dakota Gasification Company plant in Beulah, North Dakota, which converts coal to synfuels with CO_2 as a by-product, to the Weyburn oil field in Saskatchewan. The EnCana Corporation, operator of the Weyburn field, is injecting nearly 2 million tons a year of CO_2 at a depth of a little less than a mile. It plans to inject 20 million tons over the life of the project, to extract an additional 130 million barrels of oil from the field. Government and academic scientists are monitoring the reservoir intensively with seismic techniques. Weyburn is a test of the potential of old oil fields to sequester CO_2.

For two reasons, that potential is inevitably going to be limited when measured against the scale of our CO_2 problem. The first reason is a simple matter of logic. The CO_2 being injected into oil fields like Weyburn is being used to extract more oil—and when that oil is burned, it will release more CO_2 into the atmosphere than was put into the ground to extract the oil in the first place. Sequestering CO_2 at Weyburn is better than not sequestering it and getting the oil out in some other way, or than burning coal instead of the oil. But the Weyburn project and other enhanced oil recovery projects will still be a net source, rather than a sink, for CO_2. If CO_2 is injected into old oil fields without pulling out more oil, then these fields are indeed a net sink—but that brings us to the second reason why carbon sequestration in oil reservoirs is only a small start, and never the long-term solution to the CO_2 problem. There is simply nowhere near enough room in such reservoirs. The U.S. Department of Energy, which under the Bush administration has become extremely bullish on carbon sequestration, has estimated that the total CO_2 storage capacity of oil and gas fields in the United States and Canada is 82.4 billion metric tons. That's less than three years of the world's production, and a little over a decade's worth of the U.S. production of CO_2—assuming the reservoirs are filled to capacity, which they won't be.

In that same "Carbon Sequestration Atlas," released at the end of 2006, DOE identifies a much better option—deep saline aquifers like the one Statoil is exploiting under the North Sea. Such aquifers are extensive in sedimentary basins on land as well. The DOE atlas shows broad swaths of them south of the Great Lakes, along the Gulf Coast, under much of the Great Plains, and in parts of the Great Basin. Counting only the ones that lie twenty-five hundred feet deep or deeper, and thus are probably too saline to be widely used even when water is short, the DOE estimated that they have the capacity to store at least 900 billion tons of CO_2—and perhaps as much as 3.3 trillion tons. And of course such aquifers exist on all the continents and, Statoil has shown, offshore as well. The aquifer that Statoil is pumping CO_2 into at Sleipner extends under a wide stripe of the North Sea floor. Its capacity has been estimated at one hundred times the current annual emissions of all the power plants in Europe.

Since deep saline aquifers have had no economic value, they have not been explored by resource companies, and so their global capacity for storing CO_2 is highly uncertain. The IPCC report on carbon sequestration estimated it to be at least a thousand gigatons. Assuming there is something to the DOE minimum for North America of more than nine hundred gigatons, the IPCC global figure is likely to be a substantial underestimate.

One reason for the large uncertainty is that, depending on local conditions, the aquifers may store CO_2 in more than one way. They may retain it physically under a layer of caprock, as at Sleipner. But the CO_2 may also react chemically with elements of the surrounding rock and become an inert mineral—a carbonate. It is a process to be wished for, because it locks up CO_2 more or less forever, with no chance of the caprock springing a leak and CO_2 escaping back to the surface. The process happens naturally all the time, whenever rainwater laced with CO_2 contacts the right kind of rock; it is called geochemical weathering. Unfortunately, it happens at a geologically slow pace.

There are ways to speed it up, however. One way is to inject CO_2 into deep aquifers within layers of volcanic basalt. Basalt is rich in calcium silicates. When CO_2 is injected into a basalt aquifer, it first dissolves in the groundwater and becomes carbonic acid. The acid leaches calcium ions out of the silicate minerals in the surrounding rocks. The calcium reacts with the CO_2 to make carbonates.

A colleague of Broecker's at Lamont, a young Swiss scientist named Jürg Matter, has been pursuing this idea. The first place he tested it was right at Lamont. The Lamont lab is perched high above the Hudson River on the New Jersey side, on top of sheer cliffs called the Palisades. Those cliffs are made of basalt—they're the outcrop of a 750-foot-thick sheet that intruded upward into the surrounding sediments around 200 million years ago, when North

America and Europe were ripping apart to form the Atlantic Ocean, along the Mid-Atlantic Ridge. Through a six-inch pipe, Matter injected CO_2 dissolved in water into the aquifer at the base of the basalt, where it meets sedimentary rock. After a week he pumped water back out of the same well and measured the extent to which the CO_2 had reacted with the basalt minerals. The results were encouraging: they showed that the reaction released calcium from the rock fast enough to create a significant potential for storing CO_2 as carbonate.

With Broecker's help, as well as that of Lackner and Taro Takahashi at Lamont, Matter has now set his sights on a bigger arena. The entire country of Iceland is made of basalt—the volcanic outpourings of the Mid-Atlantic Ridge have been so copious that they have crested the surface and formed a forty-thousand-square-mile island. The country doesn't have a lot of trees, and the winter nights are long at sixty-five degrees north, but its situation also gives it advantages. Geothermal heat is one. Iceland is pockmarked with wells that bring hot water to the surface from depths of a mile and more. Houses in Reykjavík, the capital city, don't need furnaces; they are heated by piped-in geothermal heat, and the residents bathe in the sulfurous waters and swim in naturally heated pools. Around 18 percent of the country's electricity comes from turbines powered by geothermal steam. The rest comes from hydroelectric plants powered by the torrents rushing off of Iceland's glaciers.

Blessed as it thus is with copious clean energy, the country is not a major CO_2 emitter—most of its emissions come from cars—but it has committed itself to going completely carbon-neutral, that is, to reducing its net CO_2 emissions to zero. The government's goal is to switch cars, trucks, and the fishing fleet to using hydrogen, which it can afford to make with clean electricity, by the year 2050; the world's first hydrogen filling station is already supplying three city buses in Reykjavík. In the meantime, Iceland is interested in carbon sequestration—and with all that basalt, the country is the perfect place for it. In 2006, the president of Iceland, Ólafur Ragnar Grímsson, invited Broecker to come to the country to speak to its scientists—for a country of around three hundred thousand, Iceland is richly endowed with them too—and to try to get some kind of collaborative project going.

Grimsson himself sat in the front row at the lecture and a few days later invited Broecker to lunch at the presidential mansion, a modest white house that stands alone on a windswept peninsula across the bay from Reykjavík. Paintings by Icelandic masters decorate the walls of the elegant rooms, and at lunch Grimsson called his guests' attention to one painting in particular. It showed Reykjavík in the early twentieth century, when the air was black with coal smoke, like London's—because at that time Iceland had not yet developed its geothermal and hydroelectric resources. Instead it got most of its energy from imported British coal. Icelanders are justifiably proud of having changed that, and of the way their country has developed since. A carbon sequestration experiment, Grimsson made clear, had his firm personal support.

By the summer of 2007, Broecker and Matter had negotiated a preliminary agreement with Reykjavík Energy, the municipal agency that supplies 60 percent of the Icelandic population with hot and cold water, electricity, and sewage disposal. Reykjavík Energy has recently opened a new power plant east of the city, at a place called Hellisheidi, in the rift valley at the crest of the Mid-Atlantic Ridge. Wells drilled more than a mile into the volcanic rock tap steam at 350 degrees Celsius, but the steam is contaminated with both CO_2 and hydrogen sulfide. Hikers in the area object to the rotten-egg smell of the H_2S, and there is the carbon-neutral commitment to worry about. Klaus Lackner is helping Reykjavík Energy design a system that will separate both CO_2 and H_2S from the steam. The company has agreed to build a mile-long pipeline to carry the CO_2 from the power plant to a couple of previously drilled wells that it has made available for Matter's experiment.

In the fall of 2008, if all goes well, Matter and his colleagues will start injecting CO_2 to a depth of half a mile. There it will dissolve in water that percolates through the shattered rock between two lava flows. The acidified water will dissolve metals, mostly calcium and aluminum, from the feldspar minerals in the volcanic rock. The injected CO_2 will be tagged with radiometric carbon-14, so that its fate can be monitored at other wells downstream. As long as it remains in the form of dissolved CO_2, it might still escape back to the atmosphere through vertical cracks in the basalt—but from a depth of half a mile the leakage is expected to be small.

Once the CO_2 reacts with calcium to form ions of calcium bicarbonate, it can no longer bubble back to the atmosphere; the aquifer might carry some of it to the sea, but in the form of bicarbonate ions it will not add acidity and so will be harmless. As calcium bicarbonate accumulates in the groundwater, it will gradually precipitate out of solution to form veins of solid calcium carbonate, or calcite. Volcanic rocks laced with such white veins are proof that the process happens naturally, but the injection of concentrated CO_2 should accelerate it greatly. Perhaps the most important question for the Icelandic experiment to answer concerns the limits of the process. It is possible, for instance, that the fractures in the rock at the injection site will become clogged with calcite or with aluminum silicate clays—the CO_2 leaches aluminum silicate out of the rock too—before enough CO_2 has been disposed of to make the effort worthwhile. Theoretical calculations suggest that won't be the case. But the only way to be sure is to try it.

Even if the Icelandic experiment is a roaring success, it does not mean the world's CO_2 will be disposed of in Iceland—though that potential business opportunity is part of the reason for the country's interest. But there are vast deposits of basalt elsewhere. The Deccan Traps of west-central India, for instance, consist of more than 200,000 square miles of lava flows that are well over a mile thick. The Siberian Traps, formed by massive flows of lava 250 million years ago, are even larger, covering more than 750,000 square miles. And in the northwestern United States, in the states of Washington, Idaho, and Oregon, a similar flood basalt province lies along the Columbia River between the Cascades and the Rocky Mountains. It covers more than 60,000 square miles, and in some places the lava is more than two miles thick. DOE has estimated the storage capacity of the Columbia River basalts at between 33 and 134 gigatons of CO_2—meaninglessly precise figures, given the uncertainties, but ones that DOE says are conservative.

Klaus Lackner remains convinced that over the long term, all these forms of geological sequestration—of injecting CO_2 into natural rock formations—will not give us enough disposal capacity. In his view we will eventually need the idea Broecker once considered crazy, but no longer does—mineral sequestration. We will have to accelerate geochemical weathering by mining immense quantities of igneous rock, crushing and milling it to a fine powder, and reacting it with concentrated CO_2. If you're going to go to the trouble of mining, the best igneous rocks for the purpose are not the basalts that flowed out onto the surface, but the so-called ultramafic rocks, peridotite and serpentinite, that make up Earth's mantle. The collision of Earth's tectonic plates has brought such rocks to the surface in many places. They are much richer in magnesium silicates than basalts are, and magnesium silicates react more readily with CO_2 than do calcium silicates. Both reactions, though, actually release heat—magnesium carbonates and calcium carbonates are both lower-energy states of carbon than CO_2 is. In the thermodynamic sense, they are where the carbon in the atmosphere wants to go.

Getting it to go there fast enough isn't easy, however. Grinding the ultramafic rock to a fine powder and bathing it in CO_2 is not enough. The chemical reaction takes place between magnesium oxide or hydroxide and CO_2, but even in rock powder the magnesium is still locked up in the silicate minerals. The methods found so far to release it rapidly all involve either dissolving the silicates with an acid, which then has to be recovered, or heating the rock to hundreds of degrees and pressurizing it to many atmospheres. The cost of that, according to the IPCC study, would be about $80 per ton of CO_2—which according to Lackner is still five times too expensive to be practical. Mineral sequestration is thus probably still a few decades away. But even then we may still be waiting for solar energy and the hydrogen economy, and may thus need mineral sequestration to keep climate from deteriorating.

Mineral sequestration has one great disadvantage. To sequester a ton of CO_2, you need two tons of rock at the least. After converting the CO_2 to carbonate, you have three tons of rock to get rid of—you can shove it back in the hole you dug to get the ultramafic rock, but you'll still end up, in time, building a large mound. To get rid of the CO_2 generated by burning coal, you end up creating a new mining industry on the scale of the coal mining industry itself.

It does not have to involve blasting the tops off scenic mountains—but it will unquestionably affect the landscape. On the other hand, so would building a sixty-five-foot windmill for every man, woman, and child in the United States, and so would covering hundreds of square miles of land with solar panels, which is what it would take to provide all our energy needs with wind or solar power. There is no free lunch in solving the CO_2 problem. There is just the illusion of a free lunch we have been enjoying these past two centuries, in dumping increasing amounts of it into the atmosphere without regard to the environmental cost.

Meanwhile, mineral sequestration has two great advantages. The first is that it is unquestionably permanent. With underground injection, the disposal site would always have to be monitored for leaks. Although research and field tests so far suggest that a dangerous leak is extremely unlikely—again, natural deposits of methane and even of CO_2 itself have remained intact for millions of years—low-level chronic leakage would be inevitable if the process were ever carried out on a massive scale. The leaked CO_2 itself would add to the atmospheric burden and thus need to be captured and reinjected. Once you convert CO_2 to carbonate rock, though, you have locked it up for good—to release the carbon again you would have to heat the carbonate to hundreds of degrees, as they do in the cement industry.

The other great advantage of mineral sequestration is that its capacity is essentially limitless. There are more than enough minable ultramafics around the world to absorb all the CO_2 from all the fossil fuels we might ever burn. One particularly large deposit in oil-rich Oman is practically enough to do the job all by itself. There would be poetic justice in that.

That much space gives us room to think differently about the CO_2 problem. The problem is so daunting that all of us tend to approach it in a way that is not very ambitious. Just stabilizing CO_2 at twice the preindustrial concentration, as we have said, will be a challenging task. But that doesn't mean we should be satisfied with stabilization as a goal—any more than we would be satisfied with the cleanliness of our rivers when they no longer catch fire, or with the purity of our city air when we can see across the street again and it is not acutely dangerous to walk outside. In fact, we may well not like the climate we have with 560 parts per million of CO_2. It may not be the climate we want to bequeath to future generations; we may decide we want to clean up the mess we made by burning fossil fuels, not just stabilize the mess. In the chemistry of ultramafic rocks there is space for that dream.

SIXTEEN

Fixing Climate

In 1984, Broecker went to California to visit an old friend of his, John Nuckolls, whom he hadn't seen since the 1950s. They had met in the advanced physics courses at Wheaton College and had both ended up in graduate school at Columbia. But after a year, Nuckolls had disappeared. Broecker didn't learn until twenty-five years later that his friend had been recruited by Edward Teller to work at Lawrence Livermore National Laboratory, where Nuckolls had gone on to become the primary architect of the U.S. H-bomb arsenal. Broecker, meanwhile, had spent much of the 1960s and '70s using bomb radiocarbon and tritium to map ocean currents and the spread of fossil-fuel CO_2, never knowing that it was his friend Nuckolls who had, in a sense, injected those tracers into the atmosphere and the sea—that Nuckolls had been there, peeking out from under his lead-lined blanket, when the H-bombs went off at Bikini Atoll. It made sense, though. At Wheaton, Broecker played basketball for fun; Nuckolls built black-powder rockets. One evening Broecker was walking back from a game when he met some other students leaving the building where Nuckolls had been scheduled to give an indoor demonstration—a small rocket was supposed to zoom across the room along a wire. "Damn fool blew up the place," one of the students muttered. "They took him to the infirmary." Broecker raced over there and found his friend in a daze, his hair singed and his white shirt blackened.

When Nuckolls invited Broecker to spend a week at Livermore in 1984, a few years before he was named director of the place, the idea was that they would renew old ties by doing a little research together. At the time, Broecker was teaching his *Habitable Planet* course, was hatching his ideas about abrupt climate change and conveyor-belt shutdowns, and was thus getting ever more concerned about global warming. Nuckolls, for his part, had been fascinated by a paper by the Russian climatologist Mikhail Budyko, who had proposed that global warming could be countered by injecting sulfur dioxide into the stratosphere. The gas, Budyko said, would react with water vapor to form tiny droplets of sulfuric acid, which would reflect sunlight back to space and thereby cool the planet. Nuckolls and Broecker redid Budyko's calculations and confirmed that a doubling of atmospheric CO_2 could be roughly counterbalanced by adding 35 million tons of SO_2 to the stratosphere every year—it would have to be added every year, because it tends to precipitate out rapidly, making acid rain. The two men called Freeport Sulfur and learned that the sulfur dioxide could be had for around $20 billion; they called Boeing and were told that 747s could deliver the gas to the stratosphere for another $20 billion. They wrote a paper they called "An Insurance Policy Against a Bad CO_2 Trip"—but they never published it. To Broecker, at least, it always sounded like the sort of experiment that might easily blow up in your face and blacken your shirt.

Yet the idea has never gone away—on the contrary, as concern about the CO_2 problem has grown more urgent, interest in some sort of "geoengineering" fix has only increased. In an essay in 2006, Paul Crutzen, who won a Nobel Prize in 1995 for his work on ozone depletion, recommended reviving the SO_2-injection idea. Fossil-fuel burning, he noted, is already putting some 55 million tons a year of SO_2 into the lower atmosphere, where its cooling effect is partially offsetting global warming. As governments reduce those emissions—which cause acid rain, after all, as well as five hundred thousand premature deaths a year, according to the World Health Organization—they are making global warming worse. But in the stratosphere, SO_2 could cool the planet without killing us. So developing a way of injecting SO_2 there—Crutzen thought balloons would work better than 747s—would make sense as an "escape route against strongly increasing temperatures." The sulfur balloons should be launched, he stressed, only if all else failed.

"The chances of unexpected climate effects should not be underrated, as clearly shown by the sudden and unpredicted development of the Antarctic ozone hole," Crutzen acknowledged. Researchers at several institutions are now training their sophisticated computer models on trying to understand just what effects an SO_2 injection might have.

Crutzen is clearly not crazy—though we may hope that he will sound that way fifty years from now. Consider the cautionary case of Harrison Brown, another eminent geochemist who was Dave Keeling's postdoc adviser at Caltech. In 1953, when Keeling arrived to begin his postdoctoral work, Brown was on sabbatical, sitting on a porch in Jamaica, looking out over Ocho Rios Bay and brooding on *The Challenge of Man's Future*. One challenge he was particularly concerned with in that little book was how to provide enough food for an exploding population. The observation that plants tend to grow better when the atmosphere is rich in carbon dioxide led him to the following reflection:

> If, in some manner, the carbon-dioxide content of the atmosphere could be increased threefold, world food production might be doubled. One can visualize, on a world scale, huge carbon-dioxide generators pouring the gas into the atmosphere . . . In order to double the amount in the atmosphere, at least 500 billion tons of coal would have to be burned—an amount six times greater than that which has been consumed during all of human history. In the absence of coal . . . the carbon dioxide could be produced by heating limestone.

Brown's credentials were impeccable, his achievements substantial, and his heart in the right place. *The Challenge of Man's Future* got blurbs on its back cover from Albert Einstein ("This objective book has high value") and from William O. Douglas of the U.S. Supreme Court. Just a few years after Brown wrote it, though, Keeling would begin the measurements that showed the world it faced a fundamental challenge that Brown—like almost everyone else in his day—was completely oblivious to. From our vantage point on the Keeling curve, that passage of Brown's now sounds like something written in an antimatter universe. What we really need to visualize on a world scale in 2008 is not huge carbon dioxide generators (Brown himself decided there were more practical ways of increasing the food supply), but huge scrubbers sucking the gas *out* of the atmosphere.

There would need to be hundreds of thousands to tens of millions of them, depending on just how huge each was. "If we built a collector the size of the Great Wall of China, and it removed a hundred percent of the CO_2 that went through it, it would capture half of all the CO_2 emissions in the world," says Allen Wright. "The point is, when people think of air capture, they run away from the scale of the project. But the Great Wall of China has already been built—with animal power. That convinces me in my gut. The scale of the project is not something that should send people away."

There are more direct analogies than the Great Wall of China. As we've said, the captured CO_2 would have to be disposed of, and if the amount the world produced in a single year were spread over Manhattan, it would rise three-quarters of the way up the Empire State Building. On the other hand, if all the wastewater produced in the United States alone were spread over Manhattan, even the radio antenna on top of the Empire State would be far beneath the waves. Yet somehow in the twentieth century we managed to get our sewage problem under control. And at the same time we built an equally titanic infrastructure for extracting fossil fuels and delivering them to the billions of homes, cars, power plants, and factories that now burn them. In the twenty-first century, we can build the much smaller infrastructure needed to dispose properly of the waste from fossil-fuel burning. The task is well within our grasp.

How much would it cost? No one really knows, any more than anyone could have forecast in 1900 the cost of sewage disposal in 2000—or, for that matter, the cost of gasoline distribution in 2000 or even in 1950, when Broecker himself was pumping gas at his pop's filling station in Chicago. Klaus Lackner has estimated that CO_2 could be captured from the air and disposed of for the equivalent of around twenty-five to fifty cents per gallon of gas. The IPCC report on carbon sequestration, which did not consider air capture, cited economic modeling studies suggesting that sequestration would take off on a large scale if the price for emitting CO_2 was \$25 to \$30 per ton. Fifty dollars per ton of CO_2—equivalent to a bit more than fifty cents per gallon of gasoline—was enough, and perhaps far more than enough, to prompt Statoil to start sequestering CO_2 in the North Sea and even to develop carbon sequestration as one of its businesses. Which brings us to the one absolute certainty: no significant solution to the CO_2 problem can emerge until governments worldwide, and especially that of the United States, follow the lead of Norway and of the European Union and impose either an emissions cap or a direct tax on CO_2. Nothing can compete with a free lunch.

Governments can do more than tax carbon. They can recognize what is at stake. The greatest challenge of man's future is to provide the energy needed to lift the world's population out of poverty without imposing a cost on the planet that neither humans nor the rest of its inhabitants can bear. The answer to the challenge is in part political and in part technological, but none of the possible technologies are ready. Over the next twenty years, say, we need a massive research program to get them ready—we need to be investing heavily in research into solar energy as well as carbon capture. It does not require impossible sums of money. To do the research necessary to protect ourselves from climate change would almost certainly take less, for instance, than the \$80 billion in damages attributed to Hurricane Katrina, or the more than \$100 billion that the Bush administration requested for repairs. We are not implying, by the way, that Hurricane Katrina was caused by global warming—but we are saying that the inadequate levees in New Orleans demonstrate vividly that when the unlikely but foreseeable catastrophe finally happens, the cost of *not* having prepared for it is greater than the cost of preparing for it would have been.

While we are researching the new technologies, we will be watching how fast the ice melts in the Arctic, Greenland, and Antarctica; how fast sea level rises; how fast droughts develop in America, Australia, Africa, and the Mediterranean; how fast atmospheric CO_2 rises and the ocean acidifies in spite of our most determined efforts at conservation. Maybe climate will change slower and alternatives to fossil energy will develop faster than expected. In that case we may never need to dot the landscape with carbon dioxide scrubbers. In 2050, the very notion may sound as silly as Harrison Brown's idea of carbon dioxide generators does now. But perhaps climate will change faster than we think. In that case, by 2050 we may already be sending balloons or planes laden with sulfur dioxide into the stratosphere in an effort to avert a calamitous warming.

Since we started work on this book, the evidence that climate is changing has come in faster than ever. But so has the evidence of change in public opinion and in the political landscape. A carbon-trading system has been inaugurated in Europe. In the United States, the Supreme Court has confirmed that the government has the authority to regulate carbon dioxide if it wants to, and the next administration seems likely to want to. Al Gore has had astonishing success at making global warming the focus of a Hollywood movie and a planetwide series of rock concerts and has been rewarded with the Nobel Prize, together with the IPCC. At the concert in New Jersey, we heard a full-throated and quintessentially American anthem to the joys of the open road—but leave that aside. On balance, we are more hopeful than when we started.

Fixing climate by taking carbon dioxide back out of the atmosphere is not the same as fixing climate by putting sulfur dioxide in. It is not "geoengineering." It is much more conservative than that. Our problem with carbon dioxide is an unintended consequence of a long series of fantastic inventions—from trains, planes, and automobiles to electric light, television, and computers—that have collectively liberated the citizens of industrialized countries from want and physical labor, lengthened their lives, and enriched them tremendously. Billions of people on Earth, the kind of people who must still carry their water from a distant well or their firewood from a distant copse, remain eager for that kind of liberation. The moral stain, if there is one, lies not in our having achieved what we have by burning fossil fuels; it lies in not taking responsibility for the consequences. That's what capturing CO_2 out of the air does—

in such a way, unlike SO_2 injection, as to minimize the danger of further unintended consequences. It is not a "technical fix" that allows us to burn more fossil fuels, any more than sewage systems allow us to eat more. It is merely cleaning up after ourselves.

Most people ought to be able to agree on that moral imperative. That's one of the beauties of the air-capture idea. Because it recognizes the urgency of the problem and proposes a real solution, it should appeal to environmentalists; because it recognizes that fossil fuels aren't going away, it should appeal to oil and coal companies, which aren't going away either, and to governments such as that of the United States. It may even offer a way of moving beyond the Kyoto process and of bringing the United States, China, and India to the table. One can imagine a historic bargain. The industrial countries would agree not only to reduce their CO_2 emissions to zero, but also to take out of the atmosphere some of the CO_2 they put in over the past two centuries. In return, China and India would agree to start disposing of their own CO_2 properly by equipping all those new coal-fired power plants with the appropriate technology—which at least at first, they would be buying from the West. A new carbon-disposal industry would emerge that might become as big as the oil industry today. Somebody would get rich off it, and not necessarily Lackner and the Wright brothers. The original Wright brothers did not become aircraft-industry magnates.

Finally, as Lackner has pointed out, capturing CO_2 out of the air and storing it as carbonates may one day allow us to choose the climate we want. Our children and grandchildren, having stabilized the CO_2 level at 500 or 600 ppm, may decide, consulting their history books, that it was more agreeable at 280 ppm. No doubt our more distant descendants will choose if they can to avert the next ice age; perhaps, seeing an abrupt climate change on the horizon, they will prevent it by adjusting the carbon dioxide level in the greenhouse. By then they will no longer be burning fossil fuels, so they would have to deploy some kind of carbon dioxide generator, shades of Harrison Brown, to operate in tandem with the carbon dioxide scrubbers.

Trying to see that far into the future is crazy, of course. So too perhaps is thinking that climate could ever be controlled simply by fiddling with one dial; after a career spent trying to fathom the beast, Broecker is aware of its complexity. But thinking about how we would like our descendants to write the history of the global warming we humans caused in the twentieth and twenty-first centuries is certainly not crazy. And the most fundamental lesson to be drawn from the whole episode is that we can no longer expect Mother Earth to take care of us—the planet is ours to run, and we can't retreat from our responsibility to run it wisely. It would be good if our descendants looked back on this challenge we face now as the one that allowed us, as a species, to grow up.

Selected References

Much of the material for this book came either from Broecker's own research and recollections or from Kunzig's reporting, including interviews with other scientists. What follows is a selective list of some of the documents we consulted, which may be useful to anyone looking for more information. Two merit special mention:

The Intergovernmental Panel on Climate Change 4th Assessment Report, *Climate Change 2007*, was released in stages during 2007 and should be available in final form on www.ipcc.ch by the time our book is published. It is not fun to read but it is the single most comprehensive and authoritative overview of the climate problem.

A helpful history of scientific research into the problem is Spencer R. Weart's *Discovery of Global Warming* (Cambridge, MA: Harvard University Press, 2003). Weart's website, www.aip.org/history/climate/index.html, contains a lot of additional information. We made use of both in several places in this book.

Preface: Taming the Beast

Billings, J. S. "Sewage Disposal in Cities." *Harper's* 71:424 (September 1885): 577–84.

Broecker, W. S. *Fossil Fuel CO_2 and the Angry Climate Beast*. Palisades, NY: Eldigio Press, Lamont-Doherty Earth Observatory, 2003.

Burian, Steven J., et al. "Urban Wastewater Management in the United States: Past, Present, and Future." *Journal of Urban Technology* 7:3 (2000), 33–62.

1: Pyramid Lake, 1955

Broecker, W. S., and P. C. Orr. "Radiocarbon Chronology of Lake Lahontan and Lake Bonneville." *Geological Society of America Bulletin* 69:8 (1958), 1009–32.

Liebling, A. J. A *Reporter at Large: Dateline: Pyramid Lake, Nevada*. Reno: University of Nevada Press, 2000. Originally published in *The New Yorker* in 1955.

2: Finding Science

"College Revival Becomes Confessional Marathon." *Life*, February 20, 1950, 40–41.

3: Ice Ages and the Serb Theory

Broecker, W. S., Maurice Ewing, and Bruce C. Heezen. "Evidence for an Abrupt Change in Climate Close to 11,000 Years Ago." *American Journal of Science* 258 (June 1960), 429–48.

Escher, Siegfried. "Ignaz Venetz, Begründer der Eiszeit-Theorie, 1788–1859." *Jahrbuch der Schweizerischen Naturforschenden Gesellschaft*, 1978, 222–33.

Forel, F.-A. "Jean-Pierre Perraudin de Lourtier." *Bulletin de la Société Vaudoise des Sciences Naturelles* 35:132 (1899), 104–13.

Kelly, Meredith, Jean-François Buoncristiani, and Christian Schlüchter. "A Reconstruction of the Last Glacial Maximum (LGM) Ice-Surface Geometry in the Western Swiss Alps and Contiguous Alpine Regions in Italy and France." *Eclogae Geologicae Helvetiae* 97 (2004), 57–75.

Milankovitch, Milutin. *Durch Ferne Welten und Zeiten: Briefe eines Weltallbummlers*. Leipzig: Koehler & Amelang, 1936.

Venetz, Ignace. "Mémoire sur l'Extension des Anciens Glaciers, Renfermant Quelques Explications sur Leurs Effets Remarquables." *Nouveaux Mémoires de la Société Helvétique des Sciences Naturelles* 18 (1861).

4: Proving Milanković, Doubting Milanković

Broecker, W. S. "Absolute Dating and the Astronomical Theory of Glaciation." *Science* 151 (1966), 299–304.

———. "Climatic Change: Are we on the brink of a pronounced global warming?" *Science* 189 (1975), 460–63.

———. *The Glacial World According to Wally*. Palisades, NY: Eldigio Press, Lamont-Doherty Earth Observatory, 2002.

Denton, George H., and Terence J. Hughes, eds. *The Last Great Ice Sheets*. New York: John Wiley and Sons, 1981.

Emiliani, Cesare. "Pleistocene Temperatures." *Journal of Geology* 63 (1955), 538–78.

Hays, J. D., John Imbrie, and N. J. Shackleton. "Variations in the Earth's Orbit: Pacemaker of the Ice Ages." *Science* 194 (1976), 1121–32.

Imbrie, John, and Katherine Palmer Imbrie. *Ice Ages: Solving the Mystery*. Cambridge, MA: Harvard University Press, 1986.

Kukla, George J., Robley K. Matthews, and J. Murray Mitchell, eds. *The Present Interglacial: How and When Will It End?* Papers from the conference at Brown University, 1972. *Quaternary Research* 2:3 (1972).

5: Carbon Dioxide and the Keeling Curve

Arrhenius, Svante. *Worlds in the Making: The Evolution of the Universe*. New York: Harper and Brothers, 1908.

Callendar, G. S. "The Artificial Production of Carbon Dioxide and Its Influence on Temperature." *Quarterly Journal of the Royal Meteorological Society* 64 (1938), 223–40.

Crawford, Elisabeth. *Arrhenius: From Ionic Theory to the Greenhouse Effect*. Canton, MA: Watson Publishing International, 1996.

Keeling, Charles D. "The Concentration and Isotopic Abundances of Atmospheric

Carbon Dioxide in Rural Areas." *Geochimica et Cosmochimica Acta* 13 (1958), 322–34.

———. "Rewards and Penalties of Monitoring the Earth." *Annual Review of Energy and the Environment* 23 (1998), 25–82.

Pierrehumbert, Raymond T. "Warming the World." *Nature* 432 (2004): 677. Essay on Fourier.

Plass, Gilbert N. "Carbon Dioxide and Climate." *American Scientist* 44 (1956), 302–16.

Revelle, Roger, and Hans E. Suess. "Carbon Dioxide Exchange Between Atmosphere and Oceans and the Question of an Increase of Atmospheric CO_2 during the Past Decades." *Tellus* 9 (1957), 18–27.

6: Where the Carbon Goes

Bender, Michael L., Mark Battle, and Ralph F. Keeling. "The O_2 Balance of the Atmosphere: A Tool for Studying the Fate of Fossil-Fuel CO_2." *Annual Review of Energy and the Environment* 23 (1998), 207–23.

Broecker, Wallace S., and Edwin A. Olson. "Radiocarbon from Nuclear Tests, II." *Science* 132 (1960), 712–21.

Broecker, W. S., T. Takahashi, H. J. Simpson, and T.-H. Peng. "Fate of Fossil Fuel Carbon Dioxide and the Global Carbon Budget." *Science* 206 (1979), 409–18.

Moran, Kate, et al. "The Cenozoic Paleoenvironment of the Arctic Ocean." *Nature* 441 (2006), 601–5. See also companion papers by H. Brinkhuis et al. (606–9) and by A. Sluijs et al. (610–13).

Sabine, Christopher L., et al. "The Oceanic Sink for Anthropogenic CO_2." *Science* 305 (2004), 367–71.

Schlesinger, William H., and John Lichter. "Limited Carbon Storage in Soil and Litter of Experimental Forest Plots Under Increased Atmospheric CO_2." *Nature* 411 (2001), 466–69.

7: A Conveyor Belt in the Ocean

Broecker, Wallace S. "The Great Ocean Conveyor." *Oceanography* 4:2 (1991): 79–89.

——— "Massive Iceberg Discharges as Triggers for Global Climate Change." *Nature* 372 (1994), 421–24.

Broecker, Wallace S., and George H. Denton. "The Role of Ocean-Atmosphere Reorganizations in Glacial Cycles." *Geochimica and Cosmica Acta* 53:10 (1989), 2465–501.

———. "What Drives Glacial Cycles?" *Scientific American*, January 1990, 48–56.

Dansgaard, Willi. *Frozen Annals: Greenland Ice Sheet Research.* Odder, Denmark: Narayana Press, 2005.

Dansgaard, W., et al. "A New Greenland Deep Ice Core." *Science* 218 (1982), 1273–77.

Dansgaard, W., S. J. Johnsen, and J. Møller. "One Thousand Centuries of Climatic Record from Camp Century on the Greenland Ice Sheet." *Science* 166 (1969), 377–80.

8: Conveyor Jams, Climate Lurches

Alley, Richard B. "Abrupt Climate Change." *Scientific American*, November 2004, 62–69.

———. *The Two-Mile Time Machine: Ice Cores, Abrupt Change, and Our Future*. Princeton, NJ: Princeton University Press, 2000.

Broecker, W. S. "Does the Trigger for Abrupt Climate Change Reside in the Ocean or in the Atmosphere?" *Science* 300 (2003), 1519–22.

———. "Was the Younger Dryas Triggered by a Flood?" *Science* 312 (2006), 1146–47.

Denton, George H., Richard B. Alley, Gary C. Comer, and Wallace S. Broecker. "The Role of Seasonality in Abrupt Climate Change." *Quaternary Science Reviews* 24 (2005), 1159–82.

Denton, George H., Wallace S. Broecker, and Richard B. Alley. "The Mystery Interval 17.5 to 14.5 Kyrs Ago." *PAGES (Past Global Changes) News* 14:2 (2006), 14–16.

9: Why Worry?

Harden, Blaine. "Experts Predict Polar Bear Decline." *Washington Post*, July 7, 2005, A3.

Kolbert, Elizabeth. *Field Notes from a Catastrophe: Man, Nature, and Climate Change*. New York: Bloomsbury, 2006.

Parmesan, Camille, et al. "Poleward Shifts in Geographical Ranges of Butterfly Species Associated with Regional Warming." *Nature* 399 (1999), 579–83.

Parmesan, Camille, and Gary Yohe. "A Globally Coherent Fingerprint of Climate Change Impacts Across Natural Systems." *Nature* 421 (2003), 37–42.

Patz, Jonathan A., et al. "Impact of Regional Climate Change on Human Health." *Nature* 438 (2005), 310–17.

Stott, Peter A., D. A. Stone, and M. R. Allen. "Human Contribution to the European Heatwave of 2003." *Nature* 432 (2004), 610–14.

10: Ice Melts, Sea Level Rises

Bard, Edouard, Bruno Hamelin, and Richard G. Fairbanks. "U-Th Ages Obtained by Mass Spectrometry in Corals from Barbados: Sea Level During the Past 130,000 Years." *Nature* 346 (1990), 456–58.

Bindschadler, Robert. "Hitting the Ice Sheet Where It Hurts." *Science* 311 (2006), 1720–21.

Mercer, J. H. "West Antarctic Ice Sheet and CO_2 Greenhouse: A Threat of Disaster." *Nature* 271 (1978), 321–25.

Oppenheimer, M., and R. B. Alley. "The West Antarctic Ice Sheet and Long-Term Climate Policy." *Climatic Change* 64 (2004), 1–10.

Overpeck, Jonathan T., et al. "Paleoclimatic Evidence for Future Ice-Sheet Instability and Rapid Sea-Level Rise." *Science* 311 (2006), 1747–50.

Rahmstorf, Stefan. "A Semi-Empirical Approach to Projecting Future Sea-Level Rise." *Science* 315 (2007), 368–70.

Rignot, Eric, and Pannir Kanagaratnam. "Changes in the Velocity Structure of the Greenland Ice Sheet." *Science* 311 (2006), 986–90.

Titus, James G., et al. "Greenhouse Effect and Sea Level Rise: The Cost of Holding Back the Sea." *Coastal Management* 19 (1991), 171–204.

Velicogna, Isabella, and John Wahr. "Acceleration of Greenland Ice Mass Loss in Spring 2004." *Nature* 443 (2006), 329–31.

———. "Measurements of Time-Variable Gravity Show Mass Loss in Antarctica." *Science* 311 (2006), 1754–56.

11: Megadroughts of the Past

Cook, Edward R., et al. "Long-Term Aridity Changes in the Western United States." *Science* 306 (2004), 1015–18.

Cook, Edward R., Richard Seager, Mark A. Cane, and David W. Stahle. "North American Drought: Reconstructions, Causes, and Consequences." *Earth Science Reviews* 81 (2007), 93–134.

Held, Isaac M., and Brian J. Soden. "Robust Responses of the Hydrological Cycle to Global Warming." *Journal of Climate* 19 (2006), 5686–99.

Los Angeles Department of Water and Power. "2005 Urban Water Management Plan." www.ladwp.com.

Mote, Philip W., et al. "Declining Mountain Snowpack in Western North America." *Bulletin of the American Meteorological Society* 86 (January 2005), 39–49.

Stine, Scott. "Extreme and Persistent Drought in California and Patagonia During Mediaeval Time." *Nature* 369 (1994), 546–49.

———. "Medieval Climatic Anomaly in the Americas." In A. S. Issar and N. Brown, eds., *Water, Environment and Society in Times of Climatic Change* (New York: Springer, 1998), 43–67.

Trenberth, Kevin E., et al. "Estimates of the Global Water Budget and Its Annual Cycle Using Observational and Model Data." *Journal of Hydrometeorology* (in press).

12: The Drying of the Future

Cane, Mark A., et al. "Twentieth-Century Sea Surface Temperature Trends." *Science* 275 (1997), 957–60.

Herweijer, Celine, Richard Seager, and Edward R. Cook. "North American Droughts of the Mid-to-Late Nineteenth Century: A History, Simulation and Implication for Medieval Drought." *Holocene* 16:2 (2006), 159–71.

Hoerling, Martin, and Arun Kumar. "The Perfect Ocean for Drought." *Science* 299 (2003), 691–94.

Seager, Richard, et al. "Model Projections of an Imminent Transition to a More Arid Climate in Southwestern North America." *Science* 316 (2007), 1181–84.

13: Green Is Not Enough

Broecker, Wallace S. *How to Build a Habitable Planet*. Palisades, NY: Eldigio Press, Lamont-Doherty Earth Observatory, 1985.

Hansen, J., et al. "Dangerous Human-Made Interference with Climate: A GISS Model Study." *Atmospheric Chemistry and Physics* 7 (2007), 2287–312.

Hoffert, Martin I., et al. "Advanced Technology Paths to Global Climate Stability: Energy for a Greenhouse Planet." *Science* 298 (2002), 981–87.

Lackner, Klaus S., and Jeffrey D. Sachs. "A Robust Strategy for Sustainable Energy." *Brookings Papers on Economic Activity* 2 (2005), 215–81.

Pacala, S., and R. Socolow. "Stabilization Wedges: Solving the Climate Problem for the Next 50 Years with Current Technologies." *Science* 305 (2004), 968–72.

Raupach, Michael R., et al. "Global and Regional Drivers of Accelerating CO_2 Emissions." *Proceedings of the National Academy of Sciences* 104 (2007), 10288–93.

Robert, Paul. *The End of Oil: On the Edge of a Perilous New World*. Boston: Houghton Mifflin, 2004.

14: Scrubbing the Air

Intergovernmental Panel on Climate Change. *IPCC Special Report on Carbon Dioxide Capture and Storage*. Ed. B. Metz et al. Cambridge, UK: Cambridge University Press, 2005. www.ipcc.ch.

Lackner, Klaus S., Patrick Grimes, and Hans-Joachim Ziock. "Carbon Dioxide Extraction from Air: Is It an Option?" In *Proceedings of the 24th International Conference on Coal Utilization & Fuel Systems*, ed. B. Sakkestad, 885–96. Clearwater, FL: Coal Technology Association, 1999.

15: Disposing of Carbon

Brewer, P. G., et al. "Direct Experiments on the Ocean Disposal of Fossil Fuel CO_2." *Science* 284 (1999), 943–45.

Carbon Sequestration Atlas of the United States and Canada. Report prepared by the National Energy Technology Laboratory, U.S. Department of Energy, 2006. www.netl.doe.gov/publications/carbon_seq/refshelf.html.

Lackner, Klaus S. "Carbonate Chemistry for Sequestering Fossil Carbon." *Annual Review of Energy and the Environment* 27 (2002), 193–232.

———. "A Guide to CO_2 Sequestration." *Science* 300 (2003), 1677–78.

Langdon, C., et al. "Effect of Elevated CO_2 on the Community Metabolism of an Experimental Coral Reef." *Global Biogeochemical Cycles* 17:1 (2004), 11-1 to 11-14.

16: Fixing Climate

Brown, Harrison. *The Challenge of Man's Future*. New York: Viking Press, 1954.

Crutzen, Paul J. "Albedo Enhancement by Stratospheric Sulfur Injections: A Contribution to Resolve a Policy Dilemma?" *Climatic Change* 77 (2006), 211–19.

Morton, Oliver. "Is This What It Takes to Save the World?" *Nature* 447 (2007), 132–36.

Acknowledgments

Gary Comer, founder of Lands' End and generous supporter of climate research, urged Broecker to write a popular book on climate; Broecker said he wanted to work with a science writer; Comer offered a stipend to the writer of Broecker's choice. Thus was born this collaboration, in which the intellectual perspective came at first from Broecker and his half-century-deep experience in climate research, the writing is Kunzig's, and the book belongs to us both. Because Broecker is the subject of the book in some places as well as its coauthor, and because Kunzig interviewed and incorporated the ideas of other scientists into the narrative as well, we have chosen to refer to Broecker in the third person throughout.

We are deeply grateful to Gary Comer and sad that he did not live to see this book. For their patient efforts to make the book a good one, we would also like to thank Bill Schleicher of the Comer Science and Education Foundation; our editor, Joe Wisnovsky, and our publisher, Thomas LeBien, of Hill and Wang; our agent, Regula Noetzli; and Patty Catanzaro, who did the illustrations.

Many people gave generously of their time and energy in interviews or in commenting on the manuscript. In particular we would like to thank Broecker's longtime colleagues at the Lamont-Doherty Earth Observatory, Pierre Biscaye, Elizabeth Clark, Ed Cook, George Kukla, Jürg Matter, Dorothy Peteet, Richard Seager, Jim Simpson, and Taro Takahashi. Other essential information came from Richard Alley, Ralph Keeling, Klaus Lackner, Bill Schlesinger, Jeff Severinghaus, and Allen Wright.

Special thanks for the hospitality they showed Kunzig are due

Howard and Candace Broecker at Lake Michigan; Scott Mensing at Pyramid Lake; Scott Stine at Mono Lake; Tim Fisher and Tom Lowell at various nameless frozen lakes in Alberta; and above all George Denton, who introduced Kunzig to the Southern Alps of New Zealand, explained the tangible evidence of the Ice Age with great patience, and read the whole manuscript with a critical eye. Karen Fitzpatrick also read the manuscript as it progressed. Her comments and encouragement were essential.

We are both indebted to Broecker's assistant Moanna St. Clair for invaluable help of too many different kinds to list. Finally, Kunzig would like to thank his family and especially his children, Elizabeth and Nicholas, for all they had to put up with during the writing of the book.

Index

Page numbers in *italics* refer to illustrations.